551.46 LIV
ST

WITHDRAWN FROM STOCK

ADVANCES IN
COASTAL AND OCEAN ENGINEERING

ADVANCES IN COASTAL AND OCEAN ENGINEERING

Volume 2

Editor

Philip L.-F. Liu
Cornell University

World Scientific
Singapore • New Jersey • London • Hong Kong

Published by
World Scientific Publishing Co. Pte. Ltd.
P O Box 128, Farrer Road, Singapore 912805
USA office: Suite 1B, 1060 Main Street, River Edge, NJ 07661
UK office: 57 Shelton Street, Covent Garden, London WC2H 9HE

ADVANCES IN COASTAL AND OCEAN ENGINEERING Volume 2

Copyright © 1996 by World Scientific Publishing Co. Pte. Ltd.

All rights reserved. This book, or parts thereof, may not be reproduced in any form or by any means, electronic or mechanical, including photocopying, recording or any information storage and retrieval system now known or to be invented, without written permission from the Publisher.

For photocopying of material in this volume, please pay a copying fee through the Copyright Clearance Center, Inc., 222 Rosewood Drive, Danvers, MA 01923, USA.

ISBN 981-02-2410-9

Printed in Singapore.

PREFACE TO THE REVIEW SERIES

The rapid flow of new literature has confronted scientists and engineers of all branches with a very acute dilemma: How to keep up with new knowledge without becoming too narrowly specialized. Collections of review articles covering broad sectors of science and engineering are still the best way of sifting new knowledge critically. Comprehensive review articles written by discerning scientists and engineers not only separate lasting knowledge from the ephemeral, but also serve as guides to the literature and as stimuli to thought and to future research.

The aim of this review series is to present critical commentaries of the state-of-the-art knowledge in the field of coastal and ocean engineering. Each article will review and illuminate the development of scientific understanding of a specific engineering topic. Our plans for this series include articles on sediment transport, ocean waves, coastal and offshore structures, air-sea interactions, engineering materials, and seafloor dynamics. Critical reviews on engineering designs and practices in different countries will also be included.

P. L.-F. Liu

PREFACE TO THE SECOND VOLUME

This volume contains six papers discussing coastal processes, and physical and numerical modeling.

In the first paper, Svendsen and Putrevu give an extensive review on the state of understanding surf-zone hydrodynamics, including subjects such as wave breaking, wave-induced currents, and instability of nearshore currents and infragravity waves. They point out that although considerable progress has been made in the last three decades, there are still many areas where our understanding is far from satisfactory. The most urgent need is to develop an adequate theory for wave breaking and broken waves in the surf-zone.

One of the methods for studying the complex coastal processes is to perform laboratory experiments. However, physical models are always plagued by scale and laboratory effects because the coastal process involves many different length and time scales. In the second paper, Kamphuis presents a detailed discussion on the sources and implications of the scale and laboratory effects on physical modeling. He then discusses the differences among the long-term design model, the short-term design model, and the process model. He suggests that the future efforts should be focused on the development of composite models, linking many individual process model results by computer calculations. In other words, process models will produce appropriate coefficients or transfer functions for a complete and complex numerical model for the coastal process.

The third and the fourth papers form a two-part discussion on the mathematical modeling of the meso-tidal barrier island coasts. To understand the dynamics of coastal-inlet systems, one can either rely on empirical knowledge and construct various forms of empirical and semi-empirical models (Part I) or develop a set of mathematical models based on the physical processes (Part II). The empirical models are based on data from many inlets all over the world. Although these models do not provide the details of the dynamics, they give valuable knowledge of the equilibrium-state relationships. The semi-empirical model approach combines the empirical knowledge with basic physical principles, such as the conservation of mass. Thus, the semi-empirical models can be powerful tools for predicting the long-term dynamic behavior of tidal-inlet systems. The process-based simulation models consist of the number of modules which describe waves, current and sediment transport. Models of this type essentially use the hydrodynamic time scale (typically one tidal period). Larger time scales are reached by continuing the computations over a longer time-span. Therefore, the process-based models require significant computing efforts. de Vriend and Ribberink give a detailed review on two models, initial sedimentation/erosion-models and medium-term morphodynamics models. They also present many examples of applications.

In the fifth paper, Houston gives a brief review on different methods to mitigate beach loss caused by storms or persistent long-term erosion. He then describes, in detail, the method of beach nourishment, which is also a beach fill. This paper discusses the information that must be collected to design a beach fill and that should be monitored after the completion of the project.

The last paper of this volume shifts our attention to the design of offshore structures, such as gravity structures, floating barges and tankers. Chakrabarti discusses the effects of the uniform and shear currents on fixed and floating structures.

P. L.-F. Liu, 1995

CONTRIBUTORS

Ib A. Svendsen
Department of Civil Engineering,
University of Delaware,
Newark, Delaware 19716–6210,
USA.

Uday Putrevu
Northwest Associates,
P. O. Box 3027,
Bellevue, WA 98009
USA.

J. W. Kamphuis
Department of Civil Engineering,
Ellis Hall,
Queen's University,
Kingston, Ontario,
Canada K7L 3N6.

Huib J. de Vriend
University of Twente,
P. O. Box 217,
7500 AE Enschede,
THE NETHERLANDS.

Jan S. Ribberink
Delft Hydraulics,
P. O. Box 152,
8300 AD Emmeloord,
THE NETHERLANDS.

James R. Houston
Coastal Engineering Research Center,
U.S. Army Engineer Waterways Experiment Station,
3909 Halls Ferry Road,
Vicksburg, Missisippi 39180-6199\Zone 3,
USA.

Subrata K. Chakrabarti
Chicago Bridge & Iron Technical Services Company,
1501 North Division Street,
Plainfield, Illinois, 60544,
USA.

CONTENTS

Preface to the review series — v

Preface to the second volume — vi

Surf Zone Hydrodynamics — 1
   Ib A. Svendsen and Uday Putrevn

Physical Modelling of Coastal Processes — 79
   J. W. Kamphuis

Mathematical Modelling of Meso-Tidal Barrier Island Coasts.
Part I: Emperical and Semi-Empirical Models — 115
   Huib J. de Vriend

Mathematical Modelling of Meso-Tidal Barrier Island Coasts.
Part II: Process-Based Simulation Models — 151
   Huib J. de Vriend and Jan S. Ribberink

Beach Fill Design — 199
   James R. Houston

Shear Current and It's Effects on Fixed and Floating Structures — 231
   Subrata K. Chakrabarti

SURF-ZONE HYDRODYNAMICS

IB A. SVENDSEN and UDAY PUTREVU

In this paper we review recent developments within the topic of surf-zone dynamics. This term covers the hydrodynamics of waves and currents in the nearshore area where short period waves are breaking more or less continuously as they propagate toward the shore. Emphasis is placed on understandig the physical aspects of the flow phenomena, on illustrating the basic ideas behind recent modeling efforts and to which extent these models are able to represent the flows we actually observe in nature.

1. Introduction

Surf-zone dynamics is a highly complicated topic in hydrodynamics which deals with waves and wave-generated phenomena in the region between the breaker line on a beach and the shoreline.

When waves break on a gently sloping beach, large amounts of energy is released and turned into turbulence. As the waves continue breaking and interacting with the bottom topography, the momentum flux of the waves decreases, along with the decrease in wave height. This causes the generation of both longer-period waves and currents.

The proper analysis of the dynamics of the surf-zone requires a detailed knowledge of the breaking waves and the turbulence they create. This knowledge is not yet available. However, significant progress has been made over the last two decades, in particular, in understanding of wave-generated phenomena, such as wave setup, cross-shore, and longshore currents, and their stability, turbulence and mixing, and the generation of long-wave phenomena (surf beats, edge waves), which is also termed infragravity waves.

The significant progress made in recent decades is due to the intensive efforts of both experimental and theoretical research. Theoretical modeling was essentially initiated by the discovery of the wave radiation stress in the 1960s

(Longuet-Higgins and Stewart, 1962 and 1964) and has been ongoing with increasing intensity ever since. The collection of experimental data was mostly limited to laboratory experiments until around 1980 when the first of a series of large field experiments, the Nearshore Sediment Transport Study (NSTS) experiment on two Californian beaches, was carried out. Since then, many such collaborative field experiments, each involving an increasing number of researchers, have been conducted, first in the US, and later, on a smaller scale, in Europe and Japan.

For some time it has been a prevalent perception that the only way to truly understand the complicated processes in the surf-zone is through analysis of data from real surf-zones in the field. The models are far too simple and exclude too many of the important elements of the total picture to be able to illustrate what actually happens on a beach.

On the other hand, it has been argued that if you cannot predict even the simplest cases, such as a laboratory experiment, then how can you hope to be able to understand and dissect the highly complex picture encountered on a real beach where not only can the important parameters not be controlled but even the most extensive field measuring program will give only a sporadic glimpse of the total picture because it is practically impossible ever to measure enough?

Looking back at the development, however, it is interesting and encouraging to observe that in fact theoretical, laboratory, and field work have all contributed to new discoveries made over the last decades. The longshore currents and the surf-zone set-up were recognized early on in field data but could not be explained properly until the theoretical concept of radiation stress was firmly established. The first quantitative description of the cross-shore circulation and undertow was based on laboratory observations on a barred beach. Both surf beats and shear waves were observed in the field before they were explained theoretically. On the other hand, edge waves were known theoretically long before being observed in the field, and the 3D-vertical structure of currents and infragravity waves are theoretical predictions that, to some extent, still await full verification, as does the nonlinear mechanism of current-current and wave-current interaction. It is characteristic, however, for these and many other phenomena that a strong cross-fertilization between field and modeling efforts has taken place and no uniform pattern for progress or discovery can be identified. These two areas, supplemented by laboratory measurements, form an integral part in the history of progress towards greater understanding of the complicated nature of surf-zone dynamics.

It is our impression that, as the modeling efforts have matured to become more complete and complex and the field measurements have revealed increasingly detailed and accurate pictures of waves and currents, the dichotomy between these two approaches has been nearly wiped out. Hence, it is likely that, in the not-too-distant future, models will be able to provide additional and accurate information about details that were actually not measured in a given field experiment and also assist in the planning of new field experiments.

Today, so many contributions have been made towards our understanding of the surf-zone that it will be impossible to cover them all in one review paper. Therefore, the presentation here will, in spite of all efforts to the contrary, have to leave out, or only cover sporadically, important aspects of the discussion. In selecting the material for this paper, we have undoubtedly been influenced by our own firm association with the theoretical or modeling aspect of the issue but even there are many papers that have not been included.

In this review, we have chosen to concentrate on "recent progress" in the area of surf-zone dynamics. Thus, we have chosen not to include material that is readily available in standard text books (Phillips, 1977; Mei, 1983). A consequence of this is that the paper makes only passing references to some of the pioneering works (e.g., Longuet-Higgins and Stewart 1962 and 1964; Bowen et al., 1968; Bowen, 1969a; Thornton, 1970; Longuet-Higgins, 1970; and many others).

In addition to the books by Phillips and Mei, the reader is referred to review papers by Peregrine (1983), Battjes (1988), and Battjes et al. (1990) for recent overviews (sometimes from a different perspective) of some of the material covered here. Basic material about boundary layers in nonbreaking waves may be found in Nielsen (1992), but very little is available on boundary layers under breaking waves. Some information about wave boundary layers may also be found in texts more specifically oriented towards sediment transport such as Sleath (1984) and Fredsøe and Deigaard (1993).

The paper is organized as follows. The rest of this section is devoted to describing the basic assumptions involved in the analysis of surf-zone motions. In Sec. 2, we outline the derivation of the "short-wave-averaged" equations in the nearshore. The equations given in that section are valid for vertically nonuniform current motions and are, hence, generalized forms of the equations given by Phillips and Mei. Section 3 discusses our present knowledge of the short-wave motion in the surf-zone. A brief discussion of bottom boundary layers and bottom shear stresses is given in Sec. 4. The present understanding of steady circulation patterns (including the decay of short-waves, longshore currents, and undertow) are reviewed in Sec. 5. Sections 6 and 7 are devoted to discussing infragravity and shear waves, respectively. Section 8 discusses quasi-3D comprehensive models, and the paper concludes with a summary in Sec. 9.

Basic Assumptions

The direct approach to describing and analyzing surf-zone phenomena would require a solution of the hydrodynamical equations for the conservation of mass and momentum. Since the flow is highly turbulent due to the wave breaking and since the free surface introduces essential non-linearities, this task has not been accomplished yet.

Whereas there have been many attempts towards this goal, two major approaches have been pursued with particular success. One particularly aims at describing the pattern of currents and long ("infragravity") wave motion generated by the ("short") storm waves or swell. This approach is based on versions of the hydrodynamical equations which are averaged over the short wave period so that, in these equations, only the mean effect (over a wave period) of the short waves, such as net mass, momentum, and energy fluxes, are included in the equations.

The second approach solves the hydrodynamical equations in the time domain but only in the horizontal plane. This is made possible by approximate representations in the equations of the variations of pressure and velocity fields in the vertical direction based on the assumption that the horizontal length scale of the wave motion is much larger than the water depth. It leads to the class of descriptions that include the nonlinear shallow-water equations, Boussinesq models, and derivatives thereof.

A fundamental assumption which underlies all these efforts is the concept of a gently sloping bottom which is normally the case on littoral beaches. The gentleness of the bottom slope is used to assume that, at each location of the region, the local short-wave motion is in equilibrium with the local values of the depth, the wave height, and wave period.

It turns out that this concept of gentleness is related both to the bottom slope h_x and to the wavelength L. Analysis of the effect the bottom has on the wave motion shows that to the first order this effect is proportional to the dimensionless beach slope parameter

$$S = \frac{h_x L}{h}. \tag{1}$$

Here, L is a "local" wave length evaluated as $L = cT$, where c is the local phase velocity of the wave and T its period. Since $h_x\, L = \Delta h$ is the (first Taylor approximation to the) change in depth over one wave length, we see that S is the *relative* change in depth over that distance.

Hence, we may conclude that, if we want to be able to *neglect* the effect that a sloping bottom has on the local wave motion (i.e., to assume "locally constant depth"), we should assume conditions that satisfy the requirement that

$$S \ll 1. \tag{2}$$

This will also ensure that the assumption of no reflection of wave energy by the bottom topography is reasonable. In practice, this usually is assumed to be satisfied if $S \leq 1$ though for certain results $S < 0.3$–0.5 is probably necessary. For the larger S values, we can expect that the wave behavior will depend on the value of S. This problem, however, has not really been discussed in the literature yet.

2. The Short-Wave-Averaged Equations

2.1. *Introduction*

In this section, we give a brief account of the depth-integrated, time-averaged equations for conservation of mass and momentum. The equations are presented here for currents that are nonuniform over the depth. This is a more general form than that given for example by Phillips (1977) or Mei (1983).

Similar equations can be derived for the conservation of total energy, the conservation of oscillatory (wave) energy, and the conservation of mean (current) energy. However, nonuniform versions of the energy equations have not been presented in the literature at the present time. For the general form of the depth uniform versions of these equations, the reader is referred to the book by Phillips (1977).

In this section we also discuss the local wave-averaged equations used to determine the vertical variation of the current and long wave particle velocities in short-wave-averaged models.

2.2. *Description of the derivation*

The depth-integrated, time-averaged equations are derived from the Reynolds equations for conservation of mass:

$$\frac{\partial u_\alpha}{\partial x_\alpha} + \frac{\partial w}{\partial z} = 0 \tag{3}$$

and momentum:

$$\frac{\partial u_\beta}{\partial t} + \frac{\partial u_\alpha u_\beta}{\partial x_\alpha} + \frac{\partial u_\beta w}{\partial z} = -\frac{1}{\rho}\frac{\partial p}{\partial x_\beta} + \frac{1}{\rho}\left(\frac{\partial \tau_{\alpha\beta}}{\partial x_\alpha} + \frac{\partial \tau_{z\beta}}{\partial z}\right), \tag{4}$$

$$\frac{\partial w}{\partial t} + \frac{\partial u_\alpha w}{\partial x_\alpha} + \frac{\partial w^2}{\partial z} = -\frac{1}{\rho}\frac{\partial p}{\partial z} + \frac{1}{\rho}\left(\frac{\partial \tau_{\beta z}}{\partial x_\beta} + \frac{\partial \tau_{zz}}{\partial z}\right). \tag{5}$$

Here, x_α, z are horizontal and vertical coordinates, respectively, u_β represents the total particle velocity in the horizontal direction, w the vertical component of the total velocity, p the pressure, and $\tau_{\alpha\beta}$ the turbulent shear stresses. Figure 1 shows the definition of the geometrical quantities used throughout this paper.

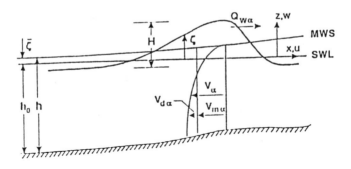

Fig. 1. Sketch defining the various geometrical quantities used in this paper.

The derivation of the short-wave-averaged equations requires the following series of operations:

- The continuity equation and the horizontal components of the momentum equations are integrated from the bottom $-h_0$ to the instantaneous free surface ζ.
- This yields terms of the form $\int \frac{\partial}{\partial t}, \int \frac{\partial}{\partial x_\alpha}$, etc., in the equations. Leibnitz's rule is used to transform those terms into terms of the form $\frac{\partial}{\partial t}\int, \frac{\partial}{\partial x_\alpha}\int$, etc.

- These steps leave a number of other terms in the equations evaluated at $-h_0$ and ζ. Invoking the exact boundary conditions at the bottom and the free surface essentially eliminates all these terms except for the normal and tangential stresses at the boundaries.
- Integrating the vertical momentum equation from the free surface to a level z gives an expression for the pressure p at level z. This can be used to eliminate the pressure from the horizontal momentum equations.
- Finally, the equations are averaged over the short-wave period. In describing the result of this process we use ─── to indicate time averaging, which means that

$$\overline{\cdot} = \frac{1}{T} \int_t^{t+T} \cdot \, dt \tag{6}$$

where T is the wave period.

2.3. The equations

Before eliminating the pressure, the depth-integrated equations of continuity and momentum can be written as:
Continuity:

$$\frac{\partial \bar{\zeta}}{\partial t} + \frac{\partial \bar{Q}_\alpha}{\partial x_\alpha} = 0 \tag{7}$$

where $\bar{Q}_\alpha$ is the total volume flux through a vertical section defined by

$$\bar{Q}_\alpha = \overline{\int_{-h_0}^{\zeta} u_\alpha \, dz} . \tag{8}$$

Momentum:

$$\rho \frac{\partial}{\partial t} \overline{\int_{-h_0}^{\zeta} u_\beta dt} + \rho \frac{\partial}{\partial x_\alpha} \overline{\int_{-h_0}^{\zeta} (u_\alpha u_\beta - \tau_{\alpha\beta}) \, dz} + \frac{\partial}{\partial x_\beta} \left(\overline{\int_{-h_0}^{\zeta} p \, dz} - \frac{1}{2} \rho g h^2 \right)$$
$$= \rho g \left(\bar{\zeta} + h_0 \right) \frac{\partial \bar{\zeta}}{\partial x_\beta} + \overline{\tau_\beta^S} - \overline{\tau_\beta^B} . \tag{9}$$

The velocity u_α represents the total instantaneous (horizontal) fluid velocity at a point, and the vertical distribution of this velocity has not yet been specified. $\overline{\tau_\beta^S}$ and $\overline{\tau_\beta^B}$ are the time-averaged surface and bottom shear stresses, respectively. As before, $\tau_{\alpha\beta}$ are the turbulent shear stresses (Reynolds stresses).

To bring the momentum equation into a more useful form, we separate the total velocity (u_α, w) into a "current" and a short-wave component by letting

$$u_\alpha(=\widehat{u_\alpha}) = V_\alpha + u_{w\alpha} \; ; \quad w(=\widehat{w}) = w_w \, . \tag{10}$$

Here, $u_{w\alpha}, w_w$ are the short-wave components which have

$$\overline{u_{w\alpha}}, \; \overline{w_w} = 0 \tag{11}$$

below wave-trough level, and V_α is the current. $\widehat{}$ represents turbulent averaging of the quantity. If the short waves are irregular, we will expect the "current" to vary with time. $V_\alpha(t)$ may then be equivalent to a long-wave particle velocity.

We also introduce the radiation stress $S'_{\alpha\beta}$ defined by

$$S'_{\alpha\beta} \equiv \overline{\int_{-h_0}^{\zeta} (\rho u_{w\alpha} u_{w\beta} + p\delta_{\alpha\beta}) \, dz} - \delta_{\alpha\beta} \frac{1}{2}\rho g h^2 \, . \tag{12}$$

The momentum equation can then be written[a] as:

$$\rho \frac{\partial}{\partial t} \bar{Q}_\beta + \rho \frac{\partial}{\partial x_\alpha} \int_{-h_0}^{\bar{\zeta}} V_\alpha V_\beta dz + \rho \frac{\partial}{\partial x_\alpha} \overline{\int_{\zeta_t}^{\zeta} u_{w\alpha} V_\beta + u_{w\beta} V_\alpha dz}$$
$$+ \rho g(\bar{\zeta} + h_0) \frac{\partial \bar{\zeta}}{\partial x_\beta} + \frac{\partial}{\partial x_\alpha} \left[S'_{\alpha\beta} - \int_{-h_0}^{\zeta} \tau_{\alpha\beta} dz \right]$$
$$- \tau^S_\beta + \tau^B_\beta = 0 \, . \tag{13}$$

In this form of the equation, we have grouped the $S'_{\alpha\beta}$ term, which is the wave contribution, and the $\tau_{\alpha\beta}$ term, which is the turbulent contribution, to the momentum flux. This grouping emphasizes the parallel mechanism behind these two terms, one caused by organized (wave) fluctuations, the other by disorganized (turbulent) fluctuations. In fact, in some texts this is further emphasized by using the same letter "S" for the two contributions:

---

[a]The derivation of this equation for the depth-uniform currents is given by Mei (1983). However, he finds it necessary to require $\nabla h_0 \ll 1$ to obtain the result because of an inappropriate use of the result for the pressure at an arbitrary (x independent) level z to determine the pressure at the (x dependent) bottom level $-h_0$.

wave radiation stress: $S'_{\alpha\beta}$,

turbulent "radiation stress": $S''^t_{\alpha\beta} = -\overline{\int_{-h_0}^{\zeta} \tau_{\alpha\beta} dz}$.

It is important at this point to emphasize that the only approximations that have been made in the derivation of these equations, apart from the usual approximations associated with fluid flow, are associated with neglecting vertical components of bottom and surface stresses (gently sloping boundaries).

Depth Uniform Currents

Equation (13) allows the currents to vary over depth, and in fact today, we not only know that nearshore currents normally do vary with the vertical location but that this depth variation is an important part of the mechanism that controls the horizontal distribution of nearshore circulation.

However, making the assumption of depth-uniform currents allows us to simplify Eq. (22) somewhat. Introducing the assumption that V_α, V_β are independent of z, (22) takes the form:

$$\rho \frac{\partial \bar{Q}_\beta}{\partial t} + \frac{\partial}{\partial x_\alpha}\left(\rho \frac{\bar{Q}_\alpha \bar{Q}_\beta}{h} + S_{\alpha\beta} - \overline{\int_{-h_0}^{\zeta} \tau_{\alpha\beta} dz}\right) = -\rho g h \frac{\partial \bar{\zeta}}{\partial x_\beta} + \tau_\beta^S - \tau_\beta^B, \quad (14)$$

which is valid for *depth-uniform currents* only. In this form of the horizontal momentum equation, the radiation stress $S_{\alpha\beta}$ is given by Eq. (23).

The form (14) is equivalent to the momentum equation used by Phillips with the addition of the turbulent stresses $\tau_{\alpha\beta}$, and the horizontal components of the surface stress τ_β^S, and the bottom shear stress τ_β^B, all of which are neglected by Phillips.

Different Forms of the Momentum Equation for Depth-Varying Currents

The momentum equation (13) is written in terms of V_α, which is the current defined in the traditional way: the net velocity at any point below wave-trough level over and above the purely oscillatory wave motion. For the general case of depth-varying currents, it is convenient to split this current into a depth-uniform and depth-varying part, and it turns out that it is relevant to consider two different ways of doing this.

One way of splitting the current is by defining V_1 such that

$$V_\alpha = \frac{\bar{Q}_\alpha}{h} + V_{1\alpha}(z). \quad (15)$$

Closer inspection shows that

$$\int_{-h_0}^{\bar{\zeta}} V_{1\alpha} dz = -Q_{w\alpha}. \tag{16}$$

If we introduce this definition into Eq. (13), the momentum equation can be written as

$$\rho \frac{\partial}{\partial t} \bar{Q}_\beta + \rho \frac{\partial}{\partial x_\alpha} \left(\frac{\bar{Q}_\alpha \bar{Q}_\beta}{h} \right) + \rho \frac{\partial}{\partial x_\alpha} \int_{-h_0}^{\bar{\zeta}} \overline{V_{1\alpha} V_{1\beta}} dz$$

$$+ \rho \frac{\partial}{\partial x_\alpha} \int_{\zeta_t}^{\zeta} \overline{u_{w\alpha} V_{1\beta} + u_{w\beta} V_{1\alpha}} dz + \rho g (\bar{\zeta} + h_0) \frac{\partial \bar{\zeta}}{\partial x_\beta}$$

$$+ \frac{\partial}{\partial x_\alpha} \left[S'_{\alpha\beta} - \int_{-h_0}^{\zeta} \tau_{\alpha\beta} dz \right] - \tau_\beta^S + \tau_\beta^B = 0. \tag{17}$$

Alternatively, the current may be divided by defining $V_{m\alpha}$ by

$$V_{m\alpha} = \frac{\bar{Q}_\alpha - Q_{w\alpha}}{h} \tag{18}$$

where

$$Q_{w\alpha} = \int_{-h_0}^{\zeta} \overline{u_{w\alpha}} \, dz \tag{19}$$

so that the depth-varying part V_d of the current can be defined by

$$V_\alpha = V_{m\alpha} + V_{d\alpha}(z). \tag{20}$$

It may be verified that

$$\int_{-h_0}^{\bar{\zeta}} V_{d\alpha} dz = 0. \tag{21}$$

Then Eq. (13) may be written as

$$\rho \frac{\partial}{\partial t} \bar{Q}_\beta + \rho \frac{\partial}{\partial x_\alpha} \left(\frac{\bar{Q}_\alpha \bar{Q}_\beta}{h} \right) + \rho \frac{\partial}{\partial x_\alpha} \int_{-h_0}^{\bar{\zeta}} \overline{V_{d\alpha} V_{d\beta}} dz$$

$$+ \rho \frac{\partial}{\partial x_\alpha} \int_{\zeta_t}^{\zeta} \overline{u_{w\alpha} V_{d\beta} + u_{w\beta} V_{d\alpha}} dz + \rho g (\bar{\zeta} + h_0) \frac{\partial \bar{\zeta}}{\partial x_\beta}$$

$$+ \frac{\partial}{\partial x_\alpha} \left[S_{\alpha\beta} - \int_{-h_0}^{\zeta} \tau_{\alpha\beta} dz \right] - \tau_\beta^S + \tau_\beta^B = 0. \tag{22}$$

where $S_{\alpha\beta}$ is a radiation stress defined by

$$S_{\alpha\beta} = S'_{\alpha\beta} - \rho\frac{Q_{w\alpha}Q_{w\beta}}{h}. \tag{23}$$

Discussion

Before we discuss the differences between the forms of Eqs. (17) and (22), it is worthwhile to discuss the role of various terms in these equations. In both the equations, the first term represents the temporal acceleration and the second term represents the convective accelerations. The $\partial\bar{\zeta}/\partial x_\beta$ term represents the pressure gradients, $S_{\alpha\beta}$ and $\tau_{\alpha\beta}$ terms the interaction between the mean flow and the short waves and turbulence, respectively. τ_β^S represents the applied surface shear stress, and τ_β^B the bottom stress. Finally, the two integral terms represent current-current and wave-current interaction terms.

We see that the two definitions of how the current is divided into a depth-uniform and a depth-varying part are closely connected to two different definitions of the radiation stress, the first of which is the definition given by Eq. (12) and the second by the expression (23). In both cases, the nonlinear interaction terms have been separated into a contribution which is equivalent to the only nonlinear term for depth-uniform currents in Eq. (14) and a set of integrals that only contain contributions from the waves and the depth-varying part of the currents.

We notice that the two forms are equivalent in the sense that the structural forms are exactly the same. The differences only occur in the definitions of the radiation stress and the way in which the current has been divided into a depth-uniform and a depth-varying part. The form (22) and the variables used in that equation correspond to a generalized version of the form introduced by Phillips (1977). Similarly, (17) resembles the momentum equation given by Mei (1983) generalized to nonuniform currents. The algebraic similarity between Eq. (12) and Mei's radiation stress is somewhat formal however, because Mei uses a different definition of $u_{w\alpha}$ by requiring $\overline{\int_{-h_0}^{\zeta} u_{w,Mei} dz} = 0$. This implies that the return current is included in his definition of the wave particle velocity. Hence, in our notation, $u_{w,Mei} = u_w - Q_w/h$. He also indicates that he neglects the $Q_{w\alpha}Q_{w\beta}/h$ term. Paradoxically, it turns out that substituting this into Mei's expression for the radiation stress we obtain an expression identical to Eq. (23). In other words, the effect of using a different wave particle velocity in the definition of the radition stress (as Mei does) is equivalent to including the $Q_{w\alpha}Q_{w\beta}/h$ term.

Comparing with Eq. (23), we see that the difference between Eqs. (17) and (22) and the equivalent equations for depth-uniform currents is represented by the current-current and wave-current interaction terms

$$\rho \frac{\partial}{\partial x_\alpha} \int_{-h_0}^{\bar{\zeta}} V_{1\alpha} V_{1\beta} dz + \rho \frac{\partial}{\partial x_\alpha} \overline{\int_{\zeta_t}^{\zeta} u_{w\alpha} V_{1\beta} + u_{w\beta} V_{1\alpha} dz} \quad \text{in Eq. (17)} \quad (24)$$

and

$$\rho \frac{\partial}{\partial x_\alpha} \int_{-h_0}^{\bar{\zeta}} V_{d\alpha} V_{d\beta} dz + \rho \frac{\partial}{\partial x_\alpha} \overline{\int_{\zeta_t}^{\zeta} u_{w\alpha} V_{d\beta} + u_{w\beta} V_{d\alpha} dz} \quad \text{in Eq. (22).} \quad (25)$$

These terms essentially represent the contribution from the depth variation of the current velocities. Little is known about the importance of these terms except that the dispersive mixing originates from these terms. This dispersive mixing appears to give important contributions to the lateral mixing for longshore currents (see Sec. 5.2 for further discussion).

Another issue needing discussion is the choice of ζ_t, the wave-trough level, as the lower limit for the integrations around the surface. This clearly is a logical choice because above that level there is only water part of the time. Thus it becomes questionable how the mean velocity (the current) should be defined in that region. It is also clear, however, from Eq. (13), that this choice does not free us from identifying what is wave and what is current above ζ_t, since it has not been possible to write the integral above that level in terms of the total velocities only and at the same time extract the wave part (which is part of the radiation stress).

Hence, it is necessary to separate the flow above the trough level into wave and current parts whatever the choice of integration limit taken. If we use $\bar{\zeta}$ as the limit for the second integral in Eq. (13), it is necessary to remember that to get an equation similar to (13) with $\bar{\zeta}$ as the lower integration limit in the second integral, it will be necessary to assume that $u_{w\alpha}$ we then define between ζ_t and $\bar{\zeta}$ satisfies Eq. (11), which means assuming $u_{w\alpha}$ defined also when there is no water above the trough. In order to generate the correct integral, the current V_α given by Eq. (10) would then have to be defined as the difference between the total (actual physical) velocity and the wave component. This implies that during the period where there is no water the current would be minus the assumed wave component.

In contrast to this, using ζ_t as the lower limit of the second integral in Eq. (13) allows us to define both the wave and the current component of the total velocity only during the period of time when there actually is water. This is our reason for choosing ζ_t as the integration limit. As mentioned earlier, however, we still have to make the separation between wave and current parts whether we choose ζ_t or $\bar{\zeta}$ as the integration limit. We also emphasize that Eqs. (17) and (22) are still exact in the same sense as Eq. (13).

2.4. *The energy equation*

The energy equation for the combined wave and current motion is needed in wave-averaged models to determine the wave height variation, and it can be derived by the same depth integration and time-averaging processes outlined for the momentum equation. In its general form, the energy equation is even more complicated than the momentum Eq. (13).

For a derivation of depth-uniform currents, reference is made to Phillips (1977). In this general form, the energy equation includes a number of terms describing the interaction between the short-wave motion and the currents/long-wave motion. These current terms, however, are usually of minor importance for the simple applications discussed here. If restricted to the wave motion only, the energy equation simply reads

$$\frac{\partial E_{f,\alpha}}{\partial x_\alpha} = \mathcal{D}. \tag{26}$$

Here, $E_{f\alpha}$ is the energy flux of the short waves in the x_α direction and $\mathcal{D}$ is the energy dissipation per unit time and the area of the bottom.

The energy flux for the waves is an abbreviation for a number of terms that emerge through the derivation of the equation. It is defined as

$$E_{f\alpha} = \overline{\int_{-h_o}^{\zeta} \left(\rho\, gz + p + \frac{1}{2}\rho(u_w^2 + v_w^2 + w_w^2) \right) u_{w\alpha} dz}. \tag{27}$$

For sine waves, Eq. (27) yields the well-known result

$$E_f = \frac{1}{16} \rho\, g\, c\, H^2 (1+G) \tag{28}$$

where $G = \frac{2kh}{sinh 2kh}$. The dissipation of energy $\mathcal{D}$ can be described by the work done by internal (turbulent) stresses, but this does not lead to a viable means of determining $\mathcal{D}$ from our present knowledge of the wave motion.

Note that in Eq. (26), energy dissipation corresponds to $\mathcal{D} < 0$. The practical evaluation of $E_{f,\alpha}$ and $\mathcal{D}$ is discussed in more detail in Sec. 3.

2.5. Wave-averaged quantities

As we have seen in Eqs. (7), (22), and (26), describing the wave generated currents and long-wave phenomena, the effects of the short waves are represented by the volume flux, Q_w, due to the wave motion, the excess momentum flux or radiation stress, $S_{\alpha\beta}$, and the energy flux, $E_{f,\alpha}$. An essential aspect of the definitions of these quantities is that they are exact in the sense that if we substitute exact short-wave expressions for the velocities and pressures in these definitions then we get the exact results for Q_w, $S_{\alpha\beta}$, and $E_{f,\alpha}$. The difficulty of course is that we do not have such exact results for the short-wave motion, particularly in the surf-zone. Therefore, it is important to realize that the approximation used for the short-wave motion is one of the major sources of inaccuracy in the prediction of nearshore circulation. An additional, important wave-averaged quantity is the energy dissipation $\mathcal{D}$ caused by the wave breaking.

To be able to predict steady nearshore circulation and long-wave phenomena from the averaged models in the surf-zone, these quantities must be expressed in terms of wave height, wave period, water depth, etc.

The radiation stress

The radiation stress is by far the most complicated of these quantities. It is worth noticing that it can be written in several useful forms. Thus, if we eliminate the pressure from Eq. (12) using the vertically integrated vertical component of the momentum equation, we get

$$S_{\alpha\beta} = \rho \overline{\int_{-h_0}^{\zeta} \left[u_{w\alpha} u_{w\beta} - \delta_{\alpha\beta} \left(w_w^2 + \widehat{w'^2} \right) \right] dz} + \frac{1}{2}\rho g \overline{\eta^2} - \rho \frac{\overline{Q_{w\alpha} Q_{w\beta}}}{h} \quad (29)$$

where w' is the vertical component of the turbulent velocity fluctuations.

In the vertical plane of the direction of wave propagation, the wave-induced particle velocities are

$$u = (u_w^2 + v_w^2)^{1/2} \quad (30)$$

$$w = w_w \quad (31)$$

and the mass flux is

$$Q_w = (Q_{wx}^2 + Q_{wy}^2)^{1/2}. \tag{32}$$

We can then define (the scalars)

$$S_m = \overline{\int_{-h_o}^{\zeta} \rho u^2 \, dz} - \rho \frac{Q_w^2}{h} \tag{33}$$

$$S_p = -\overline{\int_{-h_o}^{\zeta} \rho\, w^2 + \widehat{w'^2}\, dz} + \frac{1}{2}\rho g \overline{\eta^2} \tag{34}$$

so that

$$S_r = S_m + S_p \tag{35}$$

represents the radiation stress on a vertical surface with the normal vector in the direction of wave propagation.

The four components of $S_{\alpha\beta}$ that represent the radiation stress elements parallel and perpendicular to the x, y axes can then be written as

$$S_{\alpha\beta} = S_m\, e_{\alpha\beta} + S_p \delta_{\alpha\beta} \tag{36}$$

where

$$e_{\alpha\beta} = \begin{Bmatrix} \cos^2 \alpha_w & \sin \alpha_w \cos \alpha_w \\ \sin \alpha_w \cos \alpha_w & \sin^2 \alpha_w \end{Bmatrix}. \tag{37}$$

Hence, from the results of S_m and S_p for the radiation stress components on a surface perpendicular to the direction of wave propagation, it is possible to determine the radiation stress $S_{\alpha\beta}$ in any direction.

Notice that the negative sign in front of the $\tau_{\alpha\beta}$ term in Eq. (13) indicates the difference between the positive sign on the $u_\alpha u_\beta$ term in the traditional definition (12) for $S_{\alpha\beta}$ and the negative sign on the $u_\alpha u_\beta$ term in the definition normally adopted for $\tau_{\alpha\beta}$. This implies that the sign convention for $S_{\alpha\beta}$ and $\tau_{\alpha\beta}$ is the opposite — a point worth bearing in mind when checking direction of terms in the equations. The positive directions are shown in Fig. 2.

Dimensionless Parameters for Wave-Averaged Quantities

Without loss of generality, in the wave direction we may write the wave

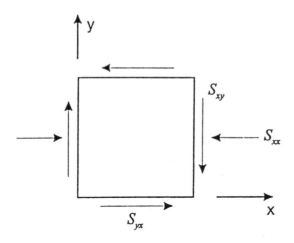

Fig. 2. Positive directions for radiation stress components. Note that the positive directions for radiation stresses are opposite to the normal positive directions for stresses.

parameters in the following way

$$Q_w = c\frac{H^2}{h}B_Q \tag{38}$$

$$S_r = \rho g H^2 P \tag{39}$$

$$E_f = \rho g c H^2 B \tag{40}$$

$$D = g\frac{H^3}{4hT}D. \tag{41}$$

Essentially, these expressions define dimensionless parameters B_Q, P, B, and D for the four wave-averaged quanties that appear in the depth-integrated, short-wave-averaged equations. In a simplified manner, one can say that the dimensional components h, H, T, and c in Eqs. (38)–(41) measure the size of the wave motion, whereas the dimensionless parameters are measures of the shape of the wave motion (understood as surface profile, velocity, and pressure fields, etc.).

One of the important questions is how accurate are the approximations (such as sine wave theory) normally used for calculating these quantities? This is discussed in Sec. 3.4.

2.6. *The local wave-averaged equations*

The local wave-averaged equations are essentially the Reynolds equations (4) in which we split the total velocity u_α into a short-wave and a current component by substituting Eq. (10) for u_α followed by a short-wave averaging. The result for the current motion can be written as (see Svendsen and Lorenz, 1989):

$$\frac{\partial U_\beta}{\partial t} + \frac{\partial U_\alpha U_\beta}{\partial x_\alpha} + \frac{\partial U_\beta W}{\partial z} + \frac{\partial (\overline{u_{w\alpha} u_{w\beta}}) - \overline{w_w^2}}{\partial x_\alpha} + \frac{\partial \overline{u_{w\beta} w_w}}{\partial z}$$
$$= -g\frac{\partial \overline{\zeta}}{\partial x_\beta} + \frac{1}{\rho}\left(\frac{\partial \tau_{\alpha\beta}}{\partial x_\alpha} + \frac{\partial \tau_{z\beta}}{\partial z}\right). \qquad (42)$$

Usually, the turbulent shear stresses $\tau_{\alpha\beta}$ in this equation are modeled by an eddy-viscosity assumption and it is also assumed that W is negligible. The resulting equation reads

$$\frac{\partial U_\beta}{\partial t} + \frac{\partial U_\alpha U_\beta}{\partial x_\alpha} + \frac{\partial (\overline{u_{w\alpha} u_{w\beta}}) - \overline{w_w^2}}{\partial x_\alpha} + \frac{\partial \overline{u_{w\beta} w_w}}{\partial z} =$$
$$-g\frac{\partial \overline{\zeta}}{\partial x_\beta} + \frac{1}{\rho}\frac{\partial}{\partial x_\alpha}\left(\nu_t\left(\frac{\partial U_\alpha}{\partial x_\beta} + \frac{\partial U_\beta}{\partial x_\alpha}\right)\right) + \frac{1}{\rho}\frac{\partial}{\partial z}\left(\nu_t\frac{\partial U_\beta}{\partial z}\right). \qquad (43)$$

Special forms of this equation have been solved for the vertical distribution of the current velocity U_β. Thus, the simplest case of steady, one-dimensional cross-shore circulation on a straight beach leads to a description of the undertow current (see Sec. 5). Also discussed in that section are other special cases, such as the vertical distribution of the longshore current on a long, straight beach and the nonlinear interaction between cross-shore and longshore currents (leading to the concept of dispersive mixing). Sec. 5.4 also includes a brief discussion of the boundary conditions used for solving Eq. (43). The time-varying velocity profiles in infragravity surf-beats (the special cross-shore form of infragravity waves in general) are discussed in Sec. 6.5. These results are all derived as solutions to special cases described by Eq. (43).

3. The Short-Wave Motion

3.1. *Introduction*

In this section, we review our present understanding of surf-zone waves. Till recently, our understanding of surf-zone waves was entirely qualitative and limited to describing patterns observed in laboratory and field experiments. These are reviewed in Sec. 3.2. As seen from the equations derived in the previous section, predictions of short-wave-averaged motions in the surf-zone require information about the integral quantities associated with surf-zone waves. Theoretical and empirical results for these quantities are discussed in Sec. 3.3. The turbulence generated by the breaking undoubtedly plays a crucial role in the dynamics of the surf-zone, and this topic is reviewed in Sec. 3.4. Finally, recent contributions to modeling breaking waves in the time domain are discussed in Sec. 3.5.

3.2. Qualitative description

Figure 3 shows a schematic of the wave motion from the breaker point to the shoreline of a gently sloping (littoral) beach. The way in which the waves break depends on the wave characteristics (deep-water wave height, wave period) and the bottom slope. The patterns range from the relatively controlled "spilling" to the violent and relatively sudden "plunging" breaker type (Galvin, 1968, 1972).

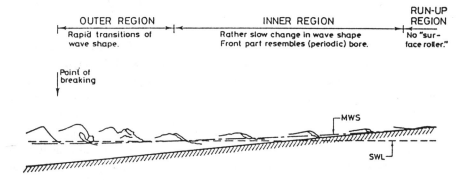

Fig. 3. A schematic description of the wave characteristics in the surf-zone. (Svendsen et al., 1978)

In any type of breaking, there will be a rapid and substantial change in the shape of the wave immediately following the initiation of breaking. This happens over a relatively short distance of 8–10 water depths after the breaker point, and this region has been termed the "outer" or "transition region" (Svendsen et al., 1978).

Shoreward of the transition region, the waves will change much more slowly. In this region, broken waves have many features in common with bores. This is the so-called "inner" or "bore region" which stretches all the way to the shore (unless the breaking occurs on a longshore bar, whereby the waves stop breaking by passing into deeper waters shoreward of the bar).

On many natural beaches, the foreshore is much steeper than the rest of the beach. In the run-up on the shore of such beaches (termed the swash zone), the wave motion often shows a different pattern from that of the rest of the surf-zone. Here, the waves sometimes turn into "surging" breakers which represent the transition stage to no-breaking or full-reflection.

If the slope becomes sufficiently steep, the waves stop breaking and full reflection occurs. The slope at which this transition occurs has not been properly studied for periodic waves, but the breaking or reflection of solitary waves on uniform slopes has been studied intensively (Synolakis 1987; Synolakis and Skjelbreia, 1993; Grilli et al., 1994). Synolakis (1987) provided experimental as well as analytical results based on Boussinesq theory. The very accurate computations for solitary waves using the Boundary Element Method (BEM) (Grilli et al., 1994) indicate that solitary waves break on a bottom slope h_x if the initial height H_0 of the wave satisfies the relation

$$\frac{H_o}{h_0} > 16.9 h_x^2 . \qquad (44)$$

where h_0 is the water depth in front of the slope. Equation (44) is an empirical result based on the computations, and Fig. 4 shows the results.

The Transition Region

The literature on the transition region is almost entirely descriptive and is based on photographic and optical methods. Basco and Yamashita (1986) give an interpretation of the flow based on such information particularly for a plunging breaker, and show how the overturning of the wave creates patterns that look chaotic but are nevertheless largely repeated from wave to wave. Similar interpretations are given by Tallent et al. (1989). Janssen (1986) has mapped the variation of the free surface in this region through high-speed video recordings of fluorescent tracers. Finally, Okayasu (1989) gives detailed measurements of the entire velocity field in the transition region from experiments using laser doppler velocimetry. The results were obtained by repeating the same experiments many times, and determining the velocity field by averaging over several waves and therefore cannot quite be regarded as a picture of the instantaneous velocity field in a particular wave.

The Bore Region

In the bore region, the information about the wave properties is also almost

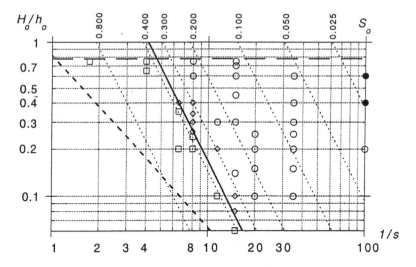

Fig. 4. Empirical breaking criterion for solitary waves on a plane slope (full line). s is the bottom slope, H_0 is the height of the wave generated at h_0. Waves to the right of the full line will break on the slope, waves to the left do not break and are fully reflected from the slope. (Grilli et al., 1994)

entirely empirical. It is only recently that predictive models of the actual wave motion have started to appear in the literature, and so far they can only predict the wave surface profiles. These models are discussed further in Sec. 3.5.

Knowledge about the waves in this region is far more quantative however, than the outer region. Among the experimental results for the bore region, it can be mentioned that Svendsen et al. (1978) found that the wave surface profiles develops a relatively steep front with a more gently sloping rear side. The shape of the rear side of the wave will change from a concave towards an almost linear variation as the waves propagate shoreward while continuing to break so that, near the shore of a gently sloping beach, the wave is close to a sawtooth shape. Figure 5 shows the tendency.

Measurements of velocity fields using laser doppler velocimetry in propagating waves have been reported by Stive (1980), Stive and Wind (1982), Nadaoka and Kondoh (1982), Nadaoka (1986), and Okayasu (1989). In all cases, however, the measurements are limited to the regions away from the crest because none of the measuring techniques available today makes it possible to measure velocities in the highly aerated region near the front of the

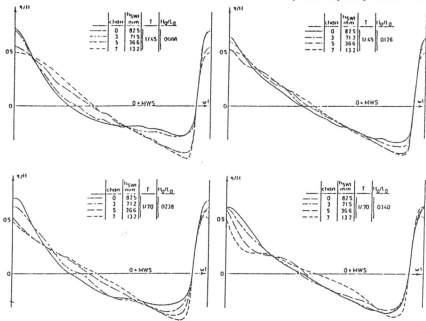

Fig. 5. The development of wave profiles in the surf-zone. (Svendsen et al., 1978)

breaker. That means wave-averaged quantities, such as radiation stresses, S_r, and energy flux, E_f, which get significant contributions from those regions, can only be determined with limited accuracy on the basis of such measurements. Stive and Wind (1982) give a detailed account of the problem. A further discussion of the available results is given in the following subsection.

3.3. Theoretical and empirical data for surf-zone waves

In most cases, linear (or "sine") wave theory has been used to calculate the wave-averaged quantities also inside the surf-zone in spite of the fact that the breaking waves are far from sinusoidal in shape and are also not of small amplitude. The wave model used by Svendsen (1984a) acknowledges that surf-zone waves are nonsinusoidal, long waves (length $\gg$ depth), and especially accounts for the fact that in breakers a volume of water (the surface roller) is carried with the wave at speed c. The situation is illustrated in Fig. 6.

Using these assumptions, it is found that, in the wave direction, the

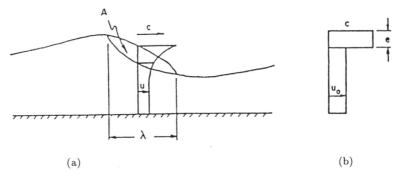

Fig. 6. An assumed vertical variation of the horizontal velocity of surf-zone waves. (Svendsen, 1984a). Note that in this model the effect of the "roller" is incorporated by assuming that the roller is carried with the wave at speed c.

radiation stress is given by

$$S_r = S_m + S_p \tag{45}$$

where

$$S_m = \rho g H^2 \left(B_0 + \frac{A}{H^2} \frac{c}{gT} \right) \tag{46}$$

$$S_p = \frac{1}{2} \rho g H^2 B_0 \tag{47}$$

$$E_f = \rho g c H^2 \left(B_0 + \frac{1}{2} \frac{A}{H^2} \frac{c}{gT} \right). \tag{48}$$

B_0 is defined as

$$B_0 = \frac{\overline{\eta^2}}{H^2}. \tag{49}$$

A is the area of the surface roller in the vertical plane. A was measured by Duncan (1981), and Svendsen (1984a) found the approximation $A/H^2 = 0.9$ constant over the surf-zone based on Duncan's data. Later, Okayasu (1989) suggested that a more accurate expression may be $A/HL = 0.06$.

The energy dissipation due to breaking is often assumed equal to the dissipation in a hydraulic jump or bore of height H. This was first suggested by Le Méhauté (1962), and has been widely used since then (e.g., Miller and Barcilon, 1978; Thornton and Guza, 1983; Svendsen 1984a; Battjes and Janssen (1978) used it in an approximate form). Then, the dimensionless dissipation D becomes

$$D = D_{\text{bore}} = \frac{h^2}{d_t d_c} \tag{50}$$

where d_t and d_c are the depths under the wave trough and wave crest, respectively (Svendsen et al., 1978). It turns out that this relationship can be expressed in terms of the wave height to water depth ratio $\frac{H}{h}$ and the ratio $\frac{\eta_c}{H}$ where η_c is the crest elevation. For most surf-zone waves, Eq. (50) gives values of $D_{\text{bore}} \sim 0.9$ (Svendsen, 1984a).

Thus, the characteristics of the wave motion used as parameters in this theory (in addition to A) are B_0, the wave phase velocity c, and η_c/H. For sine waves, $B_0 = 1/8 = 0.125$ and $\eta_c/H = 0.5$. However, all of these are physical quantities that can be measured fairly easily. Hansen (1990) analyzed original data from most of the detailed laboratory experiments available and developed empirical representations for those parameters that, in most cases, fit the data remarkably well.

Dally et al. (1985) observed that if the ratio of wave height to water depth decreases below a certain level (roughly between 0.35 and 0.40), real waves will stop breaking. They therefore asssumed that the energy dissipation at any point is given by

$$D = \frac{-K}{h}(Ec_g - (Ec_g)_s) \qquad (51)$$

where K is a dimensionless decay coefficient. The value of $(Ec_g)_s$ is chosen so that the dissipation D becomes zero when the wave height H decreases to or is below approximately 0.40 times the water depth. This approach is particularly realistic for beaches with bars or shoals where the ratio of wave height to water depth may decrease below the threshold when the waves propagate from shallower into deeper water, e.g., shoreward of a bar.

Empirical Results for Surf Zone Waves

As discussed previously, a satisfactory wave theory does not exist for surf-zone waves. In consequence, almost all of our knowledge about surf-zone waves comes from analyzing observations of breaking waves. Presently, empirical results are available for the wave celerity and some of the integral quantities (radiation stress, energy flux, and rate of energy dissipation). These results are briefly discussed below. Clearly, an accurate estimation of these quantities is crucial for proper quantitative modeling of surf-zone circulations.

The celerity of surf-zone waves has been analyzed by Svendsen et al. (1978) and Thornton and Guza (1982). Svendsen et al., analyzed the celerity of regular waves in a laboratory and found it to be somewhat higher than the shallow-water prediction ($c = \sqrt{gh}$). Thornton and Guza measured the wave celerity in a natural surf-zone. They showed that, well offshore of the surf-zone (in a water depth of 7 meters), the measured speed agreed with the predictions of linear wave theory. Inshore of this location, however, they found marked discrepancies between the measurements and linear theory predictions. In particular, they found that in the surf-zone and just offshore, the wave celerities showed weak amplitude dispersion and almost no frequency dispersion.

The energy dissipation in breaking waves has been analyzed by Svendsen et al. (1978) and Stive (1984). Both studies found that the energy dissipation in breaking waves is somewhat higher than that in a bore.

Recently, Svendsen and Putrevu (1993) analyzed a number of laboratory measurements to determine the variation of thenondimensional radiation stress (P), energy flux (B), and energy dissipation (D) inside the surf-zone. As an example, Fig. 7 shows results for the dimensionless radiation stress P for different relative bottom slopes represented by

$$S_B = \left(\frac{h_x L}{h}\right)_B \tag{52}$$

the value at the breaking point of the slope parameter mentioned earlier. h_x is the bottom slope (constant) in the experiments, $L = cT$ the wave length, and h_B the water depth at breaking. Also shown (for comparison), in Fig. 7 is the P value of 1/8 corresponding to the linear long wave theory.

Several conclusions were drawn from the results presented by Svendsen and Putrevu (1993):

1. First, the (not very surprising) conclusion that sine wave theory is inappropriate as an approximation for P (and hence B).
2. Though the variation of the wave properties such as, radiation stress, $S_{\alpha\beta}$, and energy flux, $E_{f,\alpha}$, clearly depends on the variation of the wave height (the most important parameter), the variation of the wave shape represented by P (and B) is also important for the correct prediction of radiation stress and energy flux.
3. If the breaking were almost equal to that in a bore, we would have $D \sim D_{\text{bore}}$, that is, $D \sim 1$. In most cases, the actual dissipation is substantially larger (from 50% to several hundred percent).

The Surf Similarity Parameter

The empirical results for the short-wave-averaged quantities were presented above in terms of the beach slope parameter S (introduced in Sec. 1). An

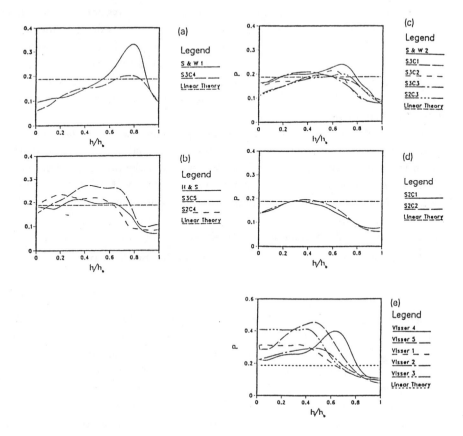

Fig. 7. Empirically determined cross-shore variations of the nondimensional radiation stress P. (Svendsen and Putrevu, 1993). The solid line in these figures represents the predictions of linear long wave theory. Curves marked S and W are based on the experiments by Stive and Wind (1982), the curves marked SxCx on are based on the experiments of Okayasu (1989), H and S is Hansen and Svendsen (1984), and Visser x is experiment x from Visser (1982). The first four figures represent four intervals of values for S_b, namely, (a) $S_b < 0.4$, (b) $0.4 < S_b < 0.5$, (c) $0.5 < S_b < 0.85$, (d) $0.85 > S_b$.

alternative parameter, the so-called Surf Similarity parameter ξ, has also been frequently used in the surf-zone. It is defined as

$$\xi = \frac{h_x}{\sqrt{H/L_0}} \tag{53}$$

where L_0 is the linear deep water wave length. This parameter was first introduced by Irribarren and Nogales (1949), and the derivation modified by Battjes (1974). The name of this parameter is somewhat misleading, however, because it is derived on the basis of standing (rather than breaking) waves on a steep beach with full reflection, and it is (somewhat heuristically it seems) assumed that the (standing) waves break at the first node from the shoreline. The depth at that node is then used as a characteristic depth which is used to determine the breaking wave height.

It is evident that this situation has little resemblance with the wave motion in an actual surf-zone where waves are propagating and breaking at a depth which is more a function of what happens seaward of the breaking point than of the distance to the shore. It is therefore surprising that this parameter has been so widely successful in classifying surf-zone conditions. It may partly be due to the fact that, on a plane beach, it has been shown that a special version of the Surf Similarity parameter, $\xi_0 = \frac{h_x}{\sqrt{H_0/L_0}}$, is related to S_b — the value of the Beach Slope parameter at the breaking point — by the expression

$$S_b = 2.30\xi_0 \qquad (54)$$

(Svendsen, 1987). Thus, since H at the break point is often close to H_0, evaluating ξ closely corresponds to evaluating the (more appropriate) parameter S_b. This could be part of the reason for the results obtained using ξ.

Whereas S_b may be the appropriate substitute for ξ, the general parameter S has another advantage in addition to being based on assumptions closely related to actual surf-zone conditions. It is a local parameter, defined at each point in the surf-zone. Hence, it can describe the situations on arbitrary beach profiles.

Thus, we have decided to use S and S_b as surf-zone parameters.

3.4. Breaker-generated turbulence

Turbulence

In a surf-zone wave, the separation of measured velocities into "wave" and "turbulent" components is not straightforward. In steady flows, the turbulent component of the velocity is usually defined using an ensemble-averaging procedure. In principle, an analogous procedure could be used to define the turbulent component in monochromatic surf-zone waves — averaging the velocity from the same phase from successive waves would lead to a quantity equivalent to an ensemble average. However, this method is far from being trivial. It turns out that even in well-controlled laboratory experiments the wave period of initially monochromatic waves does not remain constant as the waves propagate in the surf zone. This makes the identification of the phase of the wave motion problematic. As can be easily appreciated, the situation is far more complicated in a natural surf-zone where the waves are irregular. A completely satisfactory method of separating the velocity in surf-zone waves into "wave" and "turbulent" components has not been developed yet. It is important to keep this in mind while considering the results described below.

Peregrine and Svendsen (1978) experimentally found that the turbulence generated by the breaking, while initiated at the toe of the turbulent wave front, spreads downwards and continues to do so long after the breaker has passed. They speculated that the spreading mechanism is similar to that in a shear layer. Battjes and Sakai (1981) presented LDV measurements for the velocity field underneath the breaker generated by a hydrofoil positioned some distance below the water surface in a steady current. The results for the rate of variation of the turbulence indicated that the flow has more similarities to the flow in a wake. The truth is probably that the turbulence generated by wave breaking and its dispersion is different from all other turbulent phenomena.

Measurements in a steady breaker behind a hydrofoil were also made by Duncan (1981) who used photographic techniques to detemine the extent of the roller, which is the recirculating mass of water created by the highly turbulent flow down the front of the slope of the breaker. Recently, Lin and Rockwell (1994) have used the technique of Particle Image Velocimetry (PIV) to determine the entire velocity field beneath a similar breaker. They also obtained information about the vorticity field which shows that the maximum vorticity occurs in a region positioned approximately where we would expect to find the dividing steamline to the surface roller. Above that (in the "roller"), the vorticity is very weak. They concluded that the instantaneous velocity fields do not show clear signs of a vortex type of surface roller (though the present authors seem to find indications of such a roller in Fig. 2(a) in the paper). It is obvious, however, that the strong turbulent fluctuations totally dominate the weak mean flow in the roller region, and hence, the roller structure can only be expected to show up in the picture of the mean velocity field.

For completeness it is noted that not too far from the free surface the velocities under hydrofoil-generated breakers used by Battjes and Sakai (1981) and Lin and Rockwell (1994) will probably closely resemble those in a (deep water) breaking wave, whereas at some distance further down from the surface the flow is likely to be disturbed by the flow around the hydrofoil and hence be quite different from at least what we find in a surf-zone breaker.

Laboratory measurements of the distribution of turbulent intensities below the MWL in periodic waves were reported by Stive and Wind (1982), Nadaoka and Kondoh (1982), Nadaoka (1986), and in more detail by Okayasu (1989) and Ting and Kirby (1994). Thornton (1979) and George et al. (1994) reported similar measurements in the field.

Data for breaker-generated turbulence has also been provided by Hattori and Aono (1985), who found that the turbulent energy spectra have large proportions of the energy at frequencies only somewhat higher than the wave frequency indicating the existence of large scale vortices. Nadaoka (1986) and Nadaoka et al. (1989) identified a regular system of vortices with axes sloping downwards from the free surface and developing at some distance behind the front.

Battjes (1975) and later Svendsen (1987) analyzed turbulent kinetic energies under breaking waves. Battjes analyzed the value of the eddy viscosity under breaking waves based on the assumption that $\nu_t = l\sqrt{k}$ and related k, the turbulent kinetic energy, to the energy dissipated in the breaking process. Svendsen found that most of the energy is actually dissipated in the crest above the MWL. George et al. (1994) analyzed data from field experiments and compared the results for the turbulence intensities with the laboratory data. They found that while the intensity of the turbulence in the field was reduced relative to laboratory data, the characteristics of the turbulence remain the same.

The details of the highly turbulent area at the front (the so-called "roller") were analyzed by Longuet-Higgins and Turner (1974) who assumed that air entrainment played a vital part in maintaining this roller in position on the sloping front. Later results of experiments by Duncan (1981), and analysis by Svendsen and Madsen (1984), Banner (1987), and Deigaard and Fredsøe (1989) all in various ways attribute the support of the roller to turbulent shear stresses. Longuet-Higgins (1973) also analyzed the nature of the flow in the neighborhood of the toe of the roller assuming there is a separation point here. An alternative flow pattern, with a singularity in the vertical velocity gradient at the toe but continuity in the shear stress, was used in the model developed by Svendsen and Madsen (1984).

3.5. Time-domain models

Time-domain models are models that essentially solve the hydrodynamical equations in a form that can include the process of wave breaking and, hence, provide information about the phase motion. Several such models are presently in use and they are discussed below on the basis of the basic approximation used to derive the underlying equations. They are all based on long-wave assumptions, such as the Nonlinear Shallow-Water (NSW) equations or Boussinesq equations, and it has become increasingly evident that these representations are much more accurate than the nature of the underlying assumptions would lead us to believe.

Nonlinear Shallow-Water Equation Models

The nonlinear shallow-water equations are based on the assumption that the characteristic horizontal scale λ is large in comparison to the water depth h (i.e., $\mu = \frac{h}{\lambda} \ll 1$) and the wave amplitude to depth ratio $\delta = H/h$ is of order one. In Boussinesq terms, this means that the Ursell parameter $U_r = \frac{\delta}{\mu^2} = \frac{H\lambda^2}{h^3}$ is much larger than one. This leads to equations that correspond to the linear shallow-water equations (equivalent to the mild-slope equation) with additional nonlinear terms even in the first approximation. It also follows from the mentioned scale assumptions that in that approximation the pressure is hydrostatic and the horizontal velocity is uniform over depth.

This leads to the following equations which essentially are for the conservation of mass and momentum:

$$\frac{\partial \zeta}{\partial t} + \nabla_h(\mathbf{u}(h+\zeta)) = 0 \qquad (55)$$

$$\frac{\partial \mathbf{u}}{\partial t} + \mathbf{u}\nabla_h \mathbf{u} + g\nabla_h \zeta = 0 \qquad (56)$$

where $\mathbf{u}$ represents the (depth uniform) horizontal velocity vector and ∇_h is the horizontal gradient operator. Notice that in Eq. (56), the bottom friction, represented by the friction factor f, can be included by replacing the zero on the right-hand side of Eq. (56) with the term $-\frac{1}{2}\frac{f}{h}|\mathbf{u}|\mathbf{u}$.

If we disregard the bottom friction term (which is small), these equations will conserve mass and momentum and they will include no terms that represent the dissipation of energy at wave breaking. Hence, an exact solution to Eqs. (55) and (56) will also conserve energy. Due to the lack of frequency-dispersion mechanisms, they also do not have solutions of constant form: any initial wave, no matter how small, will steepen on its front side as it propagates and eventually the front will become vertical as if the wave were breaking. At that point, however, the underlying assumptions of course break down because the characteristic horizontal length (the length from trough to crest, say) is no longer large in comparison to the water depth.

The use of these equations to describe breaking waves is then based on the fact that if solved numerically by means of a Lax-Wendroff (or similar dissipative) scheme, artificial dissipation is introduced in such a way that the steepening of the front of the wave stops just before it becomes vertical. We then have a permanent-form long wave of finite amplitude for which mass and momentum are conserved. It can be shown by methods similar to the analysis of a hydraulic jump, that in such a wave energy is dissipated. On a constant depth with no change in wave form, the dissipation will then equal the dissipation in a bore of the same height (Svendsen et al., 1978), hence the relevance of the assumption that the energy dissipation in the short-wave averaged energy equation is equal to that in a bore (see Sec. 3.3).

This method of describing breaking waves has been developed extensively in the past fifteen years, starting with Hibberd and Peregrine (1979), followed by Packwood and Peregrine (1980), and Packwood (1983) which particularly discussed the effect the porous bed on sandy beaches has on the final stages of the run-up process. Watson and Peregrine (1992), Watson et al. (1992), and Barnes et al. (1994) have continued the exploration of the wave propagation using an alternative numerical method for the solution.

Other details of the wave motion, such as surface profile, particle velocities, set-up, etc., have also been analyzed using the NSW equations from just before the onset of breaking by Kobayashi et al. who first developed the computer program IBREAK (Kobayashi et al., 1989) and later the improved version RBREAK (Kobayashi and Wurjanto, 1992). They particularly explored the benefits of the method in the swash zone where the bottom slope is steep. This is particularly useful because, under those conditions, the motion is poorly covered by the gentle slope assumption underlying the wave averaged models. Much of their work deals with the steep slopes of engineering structures.

The strengths of this method include the following:

1. The method is relatively simple and robust, though the time step in the integration process needs to be kept small enough to keep the Courant number less than one. This becomes a practical problem where the water depth is shallow as in the run-up.

2. Being a time-domain model, it is also capable of describing the temporal development of random wave motion (see, e.g., Kobayashi and Wurjanto, 1992; Cox et al., 1992, 1994). It also is not restricted to a plane beach (see the same references).
3. Properly used the results such as wave surface profiles and wave heights have remarkably close resemblance with observations particularly in the swash zone (Cox et al., 1992, 1994).
4. So far these models have only been used in one-dimensional cross-shore computations. However, Eqs. (55) and (56) apply to two horizontal directions and hence can in principle describe situations with obliquely incident waves and longshore variations in the topography. With appropriate boundary conditions, the model should also be able to reproduce net flows in the swash zone, such as currents and long-wave motion, with some accuracy. There would be some doubt, though, as to the capability of these equations to correctly model that part of the radiation stress originating from the front of the breaking waves because the fronts are not accurately modeled by these equations (see below).

The weaknesses of this method are the following:

1. Since the model does not contain the mechanism that, in nature, balances the nonlinearity and resists the wave steepening (such as nonhydrostatic pressure), it cannot predict the onset of breaking. The position of breaking is determined by the distance from the offshore boundary of the computation: the waves will break at a certain distance from that boundary. Hence, the position of the offshore boundary needs to be chosen so that the model reproduces the "correct" (known) breaking point. For irregular waves, that may not always be possible for all waves in the time series.
2. The front steepens until it reaches a certain point where the (numerical) dissipation is large enough and then remains frozen. The length of the front in that situation corresponds to a few times dx. Hence, the actual form of the front is not represented by the model but depends on the choice of discretization length.
3. Having a depth-uniform velocity, the model only reproduces the depth-averaged velocity in the waves. (See point 4 above.)

Boussinesq Models

In recent years models based on Boussinesq approximations $[U_r = O(1)]$

have been extended to describe conditions similar to breaking waves. The first of these models was based on simply adding a dissipation term to the momentum equation in the Boussinesq model. Thus, Karambas et al. (1990) solve the equations

$$\frac{\partial \zeta}{\partial t} + \frac{\partial}{\partial x}\left((h+\zeta)\tilde{u}\right) = 0 \tag{57}$$

$$\frac{\partial \tilde{u}}{\partial t} + \tilde{u}\frac{\partial \tilde{u}}{\partial x} + g\frac{\partial \zeta}{\partial t} - \frac{h}{2}\frac{\partial^3 (h\tilde{u})}{\partial t \partial x^2} + \frac{h}{6}\frac{\partial^3 \tilde{u}}{\partial t \partial x^2} + g\frac{\tilde{u}|\tilde{u}|}{C^2(h+\zeta)} - \nu_t \frac{\partial^2 \tilde{u}}{\partial x^2} = 0 \tag{58}$$

in which a bottom friction term represented by the Chezy coefficient C has also been included. $\tilde{u}$ is the depth-averaged velocity; ν_t, an eddy viscosity. (See also Karambas and Koutitas (1992)).

Zelt (1991) essentially solved the same equations but on Lagrangian form, and focused on the run-up of the solitary waves. The cases with wave breaking were surging or collapsing breakers and he found that the agreement with measurements of the surface profiles of the waves was quite good though obviously the Boussinesq assumptions are not satisfied when the front of the actual wave turns vertical.

A somewhat different approach was used by Brocchini et al. (1992) and by Schäffer et al. (1992, 1993). Both these models include the effect of the surface roller in the breaking waves.

Brocchini et al. used the Serre equations. These equations represent the next order of approximation (to $O(\mu^2)$) from the NSW equations in terms of the long-wave parameter μ. Hence, they include the same dispersive terms as ordinary Boussinesq waves but assume the Ursell parameter $U_r \gg 1$. Brocchini et al. focused on the effect the roller has on the frequency dispersion term.

Schäffer et al., on the other hand, only included the effect of the roller in the nonlinear term. They solve the depth-integrated form of the Boussinsq equations

$$\frac{\partial \zeta}{\partial t} + \frac{\partial Q}{\partial x} = 0 \tag{59}$$

$$\frac{\partial Q}{\partial t} + \frac{\partial}{\partial x}\left(\frac{Q^2}{h}\right) + \frac{\partial R}{\partial x} + gh\frac{\partial \zeta}{\partial x} + \frac{h^3}{6}\frac{\partial^3}{\partial x^2 \partial t}\left(\frac{Q}{h}\right) - \frac{h^2}{2}\frac{\partial^3 Q}{\partial x^2 \partial t} = 0 \tag{60}$$

where Q is the depth-integrated horizontal velocity and R represents the only explicit effect of the surface roller. Implicit effects are included in Q however. R is calculated from a heuristically assumed velocity distribution, which includes a representation of the roller.

The results of all these models show that waves become skew and decrease in height as it is seen in the surf-zone. In the first type of model, however, the effect is achieved by including a dissipation term that essentially originates from the horizontal component of the turbulent normal stresses. This term is usually considered a small contribution to the momentum balance. The other two models (Brocchini et al. and Schäffer et al.) each use different assumptions to incorporate the roller effect. This seems likely to be the dominating effect that wave breaking has on the momentum balance. In contrast to the NSW models, the Boussinesq waves are stable and will never develop a vertical front because, as the wave height increases towards the breaking height, the crest steepens, and at this stage of the process, all Boussinesq models overestimate the dispersive effects of the surface curvature. This overstabilizes the wave and prevents vertical fronts from developing in the model. Hence, it applies to these models as well that they actually cannot predict the point at which the waves are breaking. It must be specified (e.g., from the shape of the wave surface profile together with empirical information). The velocity profiles are quadratic in the vertical coordinate. Though this is clearly a much better approximation to the actual velocity variation in waves than the depth-uniform velocity in the NSW equations, it still falls short of predicting the sharp increase in velocity near the crest of a breaking wave (van Dorn, 1976; Grilli et al., 1994).

4. Bottom Boundary Layers and Shear Stresses

Near the bed a boundary layer develops which gives rise to bottom shear stresses and locally generated turbulence.

The Bottom Shear Stress

Since the current velocity usually is weaker than the velocity amplitude in the wave motion, this boundary layer is essentially oscillatory in nature and its thickness is small in comparison to the water depth.

Under nonbreaking waves, it is important to understand the mechanisms behind the generation and diffusion of turbulence in the wave-current flow above the boundary layer because this significantly influences the current motion. In breaking waves, however, the turbulence generated in the bottom boundary layer only dominates inside that boundary layer. In the water column above the boundary layer, by far the dominating source of turbulence is the wave breaking, and the turbulence spreads downwards from the surface usually within one wave period or less rather than diffusing upwards from the bottom (see Sec. 3.4).

As a consequence of the structure of this flow, the bottom boundary layer can be expected primarily to exercise influence on the flow in the main part of the water column through the effect of the bottom shear stress. This has particular bearing on the depth-integrated, short-wave-averaged models as well as time domain models as the equations show (see, e.g., Eqs. (13) and (56)). Hence, in such models we only need to establish a relationship between the mean bottom shear stress $\overline{\tau_\alpha^B}$ and the current outside the boundary layer. Conversely, the current and wave velocities and pressure gradients determined by such models will act as forcing for the boundary layer flow.

Several such relationships have been developed through the solutions described below for the boundary layer flow. However, a practical approach introduced by Longuet-Higgins (1970) assumes that the instantaneous shear stress $\tau_\alpha^B(t)$ can be expressed in terms of a friction factor f by a relation of the form

$$\tau_\alpha^B = \frac{1}{2} \rho f \left[V_\alpha + u_{w\alpha}(t) | V_\alpha + u_{w\alpha}(t) | \right] \tag{61}$$

which can be considered a generalization of the relationship for the maximum shear stress introduced by Jonsson (1966). (Note that the summation rule does not apply to Eq. (64) because $|V_\alpha + u_{w\alpha}(t)|$ is a scalar). For weak currents and waves perpendicular to the currents, this can be simplified (after time averaging) to the following expression for the short-wave-averaged shear stress $\overline{\tau_\alpha^B}$

$$\overline{\tau_\alpha^B} = \frac{1}{\pi} \rho f u_o V_\alpha \tag{62}$$

where u_o is the bottom velocity amplitude in the waves. Liu and Dalrymple (1978) studied various cases of $\overline{\tau_\alpha^B}$ derived from Eq. (61), such as strong currents relative to the wave motion, and Svendsen and Putrevu (1990) showed that in general $\overline{\tau_\alpha^B}$ obtained from Eq. (61) can be written as

$$\overline{\tau_\alpha^B} = \frac{1}{2} \rho f u_o \{ V_\alpha \beta_1 + u_{o\alpha} \beta_2 \} \tag{63}$$

where β_1 and β_2 are functions of $u_o = |u_{o\alpha}|$ and $V_b = |V_\alpha|$ and of the angle μ between the wave and the current directions. The variable $u_{o\alpha}$ is the amplitude of the wave particle motion.

The Bottom Boundary Layer

The motion is almost always a combination of a wave and a current motion so analysis of the boundary layer flow needs to include the effect of the current. Analytical theories for wave-current interactions in bottom boundary layers have been developed by Grant and Madsen (1979), Trowbridge and Madsen (1984), Fredsøe (1983), Christoffersen and Jonsson (1985), and Davies et al. (1988), all for essentially nonbreaking wave conditions which first of all means sinusoidal or second order Stokes waves. Schäffer and Svendsen (1986) analyzed the effect of a breaker-like wave motion represented by a sawtooth time profile for the velocity above the boundary layer.

There are also numerical solutions of the boundary layer equations using one and two equation turbulent closure models. These, however, are to the authors' knowledge for situations with nonbreaking waves and are therefore not included here.

One of the questions that have been raised (Svendsen et al., 1987) is whether in fact the vortices of the breaker-induced turbulence reaching the bottom may be strong enough and sufficiently large to momentarily and locally wash away the entire bottom boundary layer.

Although important, the details of the bottom boundary layer flow is one of the topics that we have left out of this review.

5. Nearshore Circulation, Short-Wave-Averaged Models

5.1. *Introduction*

In the present section we briefly review some of the problems studied in the past twenty years. These problems have been studied using short-wave-averaged modeling, and we review the progress made in the development of these models. At the same time, the steady nearshore circulation has also been subject to substantial analysis on the basis of laboratory and, particularly, field measurements, and some of these efforts are reviewed as well in this section.

5.2. *Recent advances in steady circulations*

The Cross-Shore Wave Height and Set-Up Variation

As shown in Sec. 3, the short-wave-averaged quantities (radiation stresses, energy flux, etc.) are proportional to the square of the wave height. Briefly, one can say that in short-wave-averaged models, the solution of the energy equation will supply information on the variation of the wave height H, and hence, the mass flux and radiation stress forcing, whereas the solution of the continuity and momentum equations will provide information about water-level variations $\bar{\zeta}$ and currents V_α induced by this forcing. Therefore, the prediction of the wave height, in particular inside the surf-zone, is important for successful modeling of all the wave-generated nearshore phenomena. As also mentioned earlier, this prediction is closely linked to the correct assessment of the nondimensional parameters P for radiation stress and D for the energy dissipation due to breaking. This problem was addressed in Sec. 3.4. With these parameters known, the wave height follows from the energy equation.

The simplest possible approach, however, is to utilize the observation that on a gently, sloping beach the waves break due to the decreasing water depth, maintaining an almost constant ratio between breaking wave height and local water depth (Munk, 1949a). Thus, by this approach, the surf-zone wave height is determined as

$$H = \gamma h \tag{64}$$

where γ is assumed to be constant. Essentially, Eq. (64) replaces the energy equation (26). This assumption has been used extensively and can, in many cases, be justified by the fact that it gives qualitatively the correct type of variation of the wave height. Examples that use Eq. (64) include the first paper determining the variation of set-up $\bar{\zeta}$ in the surf-zone (Bowen et al., 1968) and those dealing with the cross-shore distribution of longshore currents (Bowen, 1969a; Thornton, 1970; Longuet-Higgins, 1970).

The simplifying assumption (64), however, is at variance with the fact that, on closer examination, the ratio of wave height to water depth in the surf-zone is not constant. This was already discovered experimentally by Horikawa and Kuo (1966). They showed that not only does the ratio decrease from an initial maximum at the breaking point, but the experiments also suggested that on a plane beach this ratio may reach a minimum near the shoreline and then increase again.

The first complete solution of the cross-shore variation of wave height and set-up, in which the energy equation was solved together with the cross-shore momentum equation instead of using Eq. (64), was given by Hwang and Divoky (1970). They used cnoidal wave theory for the short-wave-averaged quantities in the surf-zone. Later, by evaluating the short-wave-averaged parameters in the energy and momentum equations by the method described in Sec. 3.3, Svendsen (1984a) confirmed the existence of a minimum for the wave height to water depth ratio. It turns out that, by simply substituting the definitions for B and D with the energy equation (26) and rearranging the terms, this equation can be written as the following equation for the wave height to water depth ratio (Svendsen et al., 1978)

$$\left(\frac{H}{h}\right)_x = -\left(\frac{h_x}{h} + \frac{c_x}{2c} + \frac{B_x}{2B}\right)\frac{H}{h} + \frac{D}{8LB}\left(\frac{H}{h}\right)^2 \tag{65}$$

where index x represents differentiation with respect to x. Since the only assumption used to derive this equation from Eq. (26) is that the waves are periodic, this equation is well-suited to identify the various mechanisms that determine wave-height variation, particularly, in the surf-zone. The first parenthesis on the right-hand side represents the shoaling effect due to the variations in water depth which, in addition to the direct h_x-term, comes in through the change in phase velocity c and in the wave shape represented by B. The second parenthesis represents the energy dissipation (or gain through, say, wind generation if D is assumed to be > 0). It is evident from the fact that the energy term is proportional to $\left(\frac{H}{h}\right)^2$, that as the value of $\frac{H}{h}$ decreases through the surf-zone the shoaling mechanism (first parenthesis) may begin to dominate, which causes the above-mentioned minimum in $\frac{H}{h}$. It also turns out that if B and D are independent of $\frac{H}{h}$ then Eq. (65) is a Bernoulli equation which can be solved analytically (Svendsen, 1984a).

One of the observations made by Svendsen (1984a) was that, in spite of the dramatic reduction in wave height that occurs shortly after breaking, the mean water level stays horizontal for quite a distance shoreward of the break point. The only possible explanation is that the radiation stress also stays constant which, using Eq. (43), can only happen if the shape of the wave changes so that P, the nondimensional shape parameter in the radiation stress, increases in proportion to the decrease in H^2. This increase in P is reflected in the empirical results shown in Fig. 7.

In parallel with these efforts towards refining the accuracy of the prediction for regular waves, Battjes and Janssen (1978) developed the solution for energy and cross-shore momentum equations for a statistical description of random waves. In this work, the wave-height distribution is assumed to be a truncated Rayleigh distribution.

Battjes and Janssen also assumed that at any given location a certain percentage of the waves are breaking. Offshore the percentage is zero, close to the shore 100% of the waves break. Battjes and Stive (1985) calibrated this model using both laboratory and field data and showed that the model predicts the root-mean-square wave height well. The Battjes and Janssen model was refined by Thornton and Guza (1983). They analyzed field data for wave heights and showed that the Rayleigh distribution describes the random wave height variation well throughout the nearshore region. Based on their observations, Thornton and Guza propose an empirical function for the distribution of breaking waves. Thornton and Guza also show that their model predicts both the root-mean-square wave height and the distribution well. The work on the transformations of random waves in the surf zone has been continued by Dally (1990, 1992) who only assumes that the waves are initially Rayleigh distributed and calculates the ensuing change in the distribution of the wave heights as they propagate shoreward and break. The method is based on the assumption that each wave represents an isolated event for which the local wave height causes a local energy dissipation and, hence, a reduction in the height of that particular wave. The percentage Q of waves that are breaking is one of the parameters of the model that is still being investigated.

At present, it is found that the wave-height-setup models are quite accurate for the prediction of the wave height variation, in particular, such models that include empirical constants which have actually been calibrated to predict the experimental results for the wave height. The major inaccuracy usually occurs in the prediction of the setup, even if the delay after breaking is included artificially. The reason seems to be that the assumptions made about the value of P, the nondimensional radiation stress (in particular the use of sine wave theory), are too inaccurate. This is unfortunate because, as Eq. (36) shows the radiation stress in any direction can be determined from the radiation stress of plane waves, and as (13) shows, the gradient of the general radiation stress is the major forcing for currents and infragravity motion in the surf-zone. Hence the inaccuracy of these models in predicting the cross-shore set-up on a simple, long straight coast really represents a similar inability to predict the proper forcing in general nearshore circulation problems (Svendsen and Putrevu, 1993).

Longshore Currents

The first solutions to the cross-shore variation of the longshore current pattern were given by Bowen (1969a), Thornton (1970), and Longuet-Higgins (1970). These early works clearly demonstrated that it was necessary to include lateral mixing to properly model the cross-shore distribution of longshore currents. As a result, much effort was devoted to understanding the mechanisms responsible for this mixing (early discussions may be found in Inman *et al.*, 1971), Bowen and Inman (1974) and Battjes (1975)).

In most investigations, such mixing inside the surf-zone was attributed to the strong turbulence generated by wave breaking in that region (see, e.g., the discussion in Mei (1983) pp. 484-485). The works of Bowen, Thornton, and Longuet-Higgins also suggested that it was necessary to assume a level of mixing outside the surf-zone that was of the same order as the mixing inside the surf-zone. However, no explanation was forthcoming about the source of mixing outside the surf-zone as observations (Harris et al., 1963; Inman et al., 1971; Nadaoka and Kondoh, 1982) showed that the turbulence outside the surf-zone was very weak.

Nearshore Mixing

In the field, the random nature of the incident wave field leads to a time variation of the point at which each individual wave is breaking. This corresponds to variations in the width of the forcing region for longshore currents. Thornton and Guza (1986) demonstrated that these variations, when averaged over some time, provide an effect that is similar to mixing. They also demonstrated that when the time variation of the break point is taken into account, the additional smoothing of the longshore current profile, by including the turbulent mixing, is minor. Thus, Thornton and Guza suggest that the smoothing of the longshore current profile in the field may be due to the random nature of the incident wave field. While this mechanism is clearly important, it does not provide an explanation for the high level of mixing found in laboratory measurements of longshore currents generated by regular waves (e.g., Visser, 1984).

Oltman-Shay et al. (1989) identified temporal oscillations of longshore currents in the field. These oscillations (further discussed in Sec. 7) may be interpreted as a shear instability of the longshore current (Bowen and Holman, 1989). Such instabilities contribute a mixing mechanism which, under certain conditions, may be strong enough to account for the required mixing (Dodd and Thornton, 1990; Putrevu and Svendsen, 1992a; Church et al., 1994). However, theoretical estimates (Putrevu and Svendsen 1992a) as well as laboratory measurements (Reniers et al., 1994) suggest that it is unlikely that these instabilities are developed under laboratory conditions involving plane beaches. Hence, they cannot account for the mixing observed in the laboratory experiments even though they clearly represent an important mechanism in the field.

Putrevu and Svendsen (1992b) and Svendsen and Putrevu (1994a) showed that the vertical nonuniformity of the current profiles leads to lateral mixing through a momentum dispersion mechanism. This mechanism is analogous to the dispersion of pollutants in a shear flow first discovered by Taylor (1954) and expanded on (among others) by Elder (1959) and Fischer (1978). Given below is a brief account of the mechanism considered by Svendsen and Putrevu.

On a long, straight coast, the cross-shore distribution of a longshore uniform, steady longshore currents is governed by

$$\frac{d}{dx}\left(\int_{-h_0}^{\bar{\zeta}} UV dz + \overline{\int_{\zeta_t}^{\zeta} u_w V + v_w U dz}\right) + \frac{1}{\rho}\frac{d}{dx}\left[S_{xy} - \overline{\int_{-h_0}^{\zeta} \tau_{xy} dz}\right] + \tau_y^B = 0 \tag{66}$$

where U is the undertow, V the longshore current, and τ_{xy} the Reynolds' stress which is usually parameterized as

$$\tau_{xy} = \rho \nu_t \frac{dV_m}{dx} \tag{67}$$

where V_m is the depth-averaged longshore current.

The first two terms in the first parenthesis of Eq. (66) arise from the vertical nonuniformity of the currents and are usually neglected in longshore current calculations (see, e.g., Mei (1983) Eq. (5.3), p. 471). The last term in the first parenthesis is either neglected (Mei) or absorbed in the definition of the radiation stress (Phillips, 1977).

The $\int_{-h_0}^{\bar{\zeta}} UV dz$ term in Eq. (66) represents the cross-shore transport of longshore momentum by the cross-shore current and the $\overline{\int_{\zeta_t}^{\zeta} u_w V}$ term represents the net cross-shore transport of longshore momentum by the waves. Hence, these terms together represent the total cross-shore transport of longshore momentum by the waves and the currents. It is easily verified that in a situation with no net cross-shore mass flux, these terms oppose and cancel each other in the special case of depth-uniform longshore currents.

Svendsen and Putrevu showed that the paranthetic term in Eq. (66) may be written as

$$\int_{-h_0}^{\bar{\zeta}} UV dz + \overline{\int_{\zeta_t}^{\zeta} u_w V + v_w U dz} = -D_c h \frac{dV_m}{dx} + F_1 V_m + F_2 \tag{68}$$

where D_c (the dispersion coefficient) is given by

$$D_c = \frac{1}{h}\int_{-h_0}^{\bar{\zeta}} U \int_z^{\bar{\zeta}} \frac{1}{\nu_t}\int_{-h_0}^{z} U\, dz\, dz\, dz. \tag{69}$$

F_1 and F_2 are similar, but less important, coefficients determined by the vertical structure of the currents. Substituting Eqs. (67) and (68) into Eq. (66) shows that the D_c term adds to the lateral mixing caused by the turbulent Reynolds stresses. It turns out that for typical surf-zone conditions, $D_c \gg \nu_t$. For example, Fig. 8 shows typical cross-shore variations of the dispersion and turbulent eddy viscosity coefficients. It is clear from this figure that $D_c \gg \nu_t$, which means that lateral mixing is totally dominated by the dispersion mechanism in the nearshore.

The calculations described by Svendsen and Putrevu lead to the conclusion that the lateral structure of nearshore currents is controlled by the vertical structure of those currents. While this result is somewhat surprising, analogues of this result for the lateral spreading of contaminants in shear flows are well known (see, e.g., Fischer *et al.* (1979) Ch. 4).

One of the important consequences of this result is that it unifies (at least for the a longshore uniform situation) the estimates of the magnitude of the eddy viscosity in the surf zone required for predictions of different phenomena, such as the crosshore distribution of longshore currents and the vertical distribution of both cross-shore circulation and the longshoe current velocities. It furthermore suggests that the major part of the lateral mixing may be due to predictable nonlinear interactions, rather than turbulent effects which, in the models of today, are only described empirically or semi-empirically through higher-order closure models.

A Current Problem in Longshore Current Modeling

Recently, the prediction of longshore currents on barred beaches has become a topic of intense research interest. This motivation comes from field observations which show that the strongest longshore current on beaches with sand bars often occurs in the trough between the bar and the shoreline (Bruun, 1963; Allender and Ditmars, 1981; Greenwood and Sherman, 1986; Church and Thornton, 1993; Smith *et al.*, 1993; Kuriyama and Ozaki, 1993). Efforts to reproduce this pattern on the basis of theoretical models that assume longshore uniformity result in longshore velocity profiles with two maxima, one over or close to the crest of the bar and one very close to the shoreline (see e.g. Church and Thornton 1993; Smith *et al.*, 1993; Allender *et al.*, 1978; Ebersole and Dalrymple, 1980; and Larson and Kraus, 1991). Various modifications based on increasing the mixing in the trough behind the bar or adding a turbulent transport equation have been unable to significantly change this main

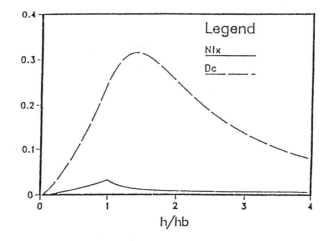

Fig. 8. Cross-shore variations of the dispersion coefficient D_c and the turbulent eddy viscosity ν_t. (Putrevu and Svendsen, 1992b)

result (see also Church et al. (1994)). It is important to emphasize, however, that the dispersive process of mixing cannot create a maximum velocity in the middle of a steady flow.

There have been a number of informal suggestions that the maximum in the trough could be attributed to longshore pressure gradients. Church and Thornton (1993) discuss this possibility. Based on the high level of correlation between the observed direction of the longshore current in the trough of the bar and the changes in the wave incidence quadrant, they argue that longshore pressure gradients are unlikely to be the cause of the observed longshore current maximum in the trough of the bar. On the other hand, recent laboratory measurements by Reniers et al. (1994) indicate that, under conditions of longshore uniformity, the maximum longshore current occurs over the bar crest as predicted by the models, not in the trough of the bar. This observation suggests that the longshore current maximum in the trough of the bar could be attributed to longshore variations (including longshore gradients in the mean water level) that were not adequately resolved by the instrument arrays in the field experiments. The occurrence of wave breaking in the trough behind the bar (Lippmann and Holman 1992) and the occurence of shear waves are additional features, the effects of which have yet to be fully analyzed in this context.

Longshore Nonuniform Longshore Currents

Variations in the bottom topography will usually be associated with similar variations in the pressure (due to mean water level changes). Longshore variations of the bottom topography will therefore lead to pressure gradients in the longshore direction. The effect that longshore pressure gradients (caused either by topographic variations in the surf-zone or by longshore variations in the wave height and angle at breaking) have on longshore current distributions has received very little attention even though it has been clearly demonstrated that relatively small longshore pressure gradients can drive strong longshore currents (Dalrymple, 1978; Wu et al., 1985). For example, Wu et al., compared the predictions of a 2D surf-zone circulation model with data obtained during the NSTS experiment. Their computations showed that a topography with even a relatively noticeable longshore variability will create substantial longshore changes in the current (Fig. 9). Recently, Putrevu et al. (1995) expanded Mei and Liu's (1977) work to study the effect of longshore topographic variability on longshore current predictions under simplified circumstances. The results confirm that longshore pressure gradients can alter the longshore currents substantially.

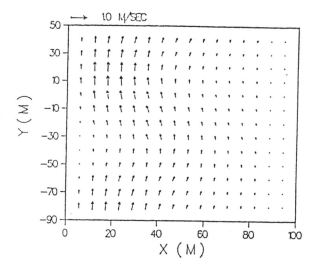

Fig. 9. Calculated variations of the nearshore currents over an a longshore varying bottom topography. (Wu et al., 1985)

5.3. *Depth-averaged, 2D-horizontal models, nearshore circulation*

Nearshore circulation refers to the current patterns generated by waves, wind, and tidal motion. It has long since been recognized as a major source of coastal change. The depth-averaged circulation in the horizontal plane can be described by wave-averaged models and the first of these was developed by Noda *et al.* (1974). Such models traditionally solved the wave-averaged equations (7) and (14) assuming that the short-wave-averaged velocities are uniform over depth. The bottom boundary layer only influences the flow through the short-wave-averaged shear stress and is usually related to the mean velocity through a friction factor as described in Sec. 4. In Eq. (14) the turbulent shear stresses $\tau_{\alpha\beta}$ along the vertical sides of the water coloumn are normally modeled by means of an (lateral) eddy viscosity.

Noda *et al.* (1974) cast these equations in a special form by introducing a stream function ψ, but in most models the equations are solved directly for the physical variables $\bar{\zeta}$ and Q_α. The primary forcing in the equations is the variation of the radiation stresses and they are usually determined from the wave-height variation using sine wave theory. As can be seen from Eq. (14), however, it is straightforward to include the effect of wind stresses as well.

Inside the surf-zone, the wave height is invariably determined from Eq. (64) whereas the wave height outside the surf-zone is often determined using various techniques for calculating the refraction pattern. This allows for application to arbitrary bottom topography. Examples include Noda *et al.* (1974) who use the ray tracing method of Munk and Arthur (1952), and Ebersole and Dalrymple (1980), Wu and Liu (1985), and Watanabe (1982) who all use the fact that the wave number field is irrotational. In some cases, the models are limited to long, straight beaches with longshore uniformity for which Snell's law applies, which means the wave heights and directions are readily known. The model presented by Wind and Vreugdenhil (1986) is special in that it uses a two-equation turbulent closure model to determine the eddy viscosity and the dissipation of turbulent kinetic energy. The wave heights are given by Eq. (64) however, and the dissipation of wave energy is then calculated from the wave-averaged-energy equation. This dissipation is used in the turbulent closure equations to determine the eddy viscosity. The eddy viscosity is then used to determine the lateral mixing. The accuracy of this approach is unknown.

Some of the models are used for problems with a periodic bottom topography, in connection with periodicity boundary conditions at the upstream and downstream boundaries (Noda, Ebersole and Dalrymple), some analyze the circulation in a closed basin with various geometric features (Wu and Liu, 1985; Wind and Vreugdenhil, 1986; Watanabe, 1982) with similarity to the circulation patterns associated with rip currents.

In general, these models have only been used sporadically to analyze the many problems of coastal circulation which they, in spite of their simplified form, are actually capable of describing.

One of the major limitations at the moment is probably the absence of an appropriate wave driver model (see Sec. 8 for further discussion).

Rip Currents

Depth-averaged, 2D, horizontal models have also been used to study rip currents. These strong, seaward-oriented, jet-like flows are sometimes found emanating from the surf-zone in nature (e.g., Shepard *et al.*, 1941; Shepard and Inman, 1950; Harris *et al.*, 1963; Bowen and Inman, 1969; Inman *et al.*, 1971; Sonu, 1972; Dalrymple and Lozano, 1978; to mention a few). These seaward-oriented currents, which are often periodic in the longshore direction (e.g., Fig. 10), are called rip currents and are a particular source of concern to swimmers. Observations have shown that these narrow seaward-oriented jets disintegrate outside the surf-zone. Most of the field observations further show that the rip currents occur at locations where the wave height is the lowest. The reader is referred to Hammack *et al.* (1991) for a set of laboratory observations that provide a description of the characteristics of rip currents. For somewhat more detailed overviews of rip currents than that presented below, the reader is referred to Dalrymple (1978) and Tang and Dalrymple (1988).

A number of possible mechanisms have been proposed for the generation of rip currents. According to Tang and Dalrymple (1988), the generation models can be broadly classified into three categories:

1. a longshore nonuniformities,
2. wave-wave interactions, and
3. instability mechanisms

In the first class of models, it is hypothesized that rip currents are generated by longshore variations of the wave height or bottom topography, which create longshore variations of the radiation stresses that in turn drive the rip currents. Bowen (1969b) showed that longshore variations of the breaker height force currents that flow seaward at locations where the wave heights are lowest. Noda (1974) used his circulation model to analyze similar situations and claimed favorable agreement with Sonu's (1972) observations. Mei and Liu (1977)

Fig. 10. An example observation of rip currents. (Inman et al., 1971)

developed analytical results of the circulation in the presence of a longshore perturbation of a plane beach. For simplicity, the variations were assumed to be sinusoidal. Their results showed that the direction of the circulation was controlled by the ratio of the surf-zone width to the a longshore wave length of the bottom topography. Dalrymple (1978) considered the case of normally incident, nonbreaking waves propagating over a longshore bar with periodic rip-channel openings. He assumed, on a heuristic basis, that the 2D-horizontal flow pattern could be treated as a combination of two 1D flows: the shoreward cross-shore flow of water over the bar caused by the nonbreaking waves, and the longshore flow in the trough that was fed by the cross-shore flow and hence had an increasing discharge as it approached the rip channel. His analysis indicated that gaps in a longshore bars do produce rip currents.

The presence of lateral boundaries (e.g., breakwaters) introduces another type of longshore nonuniformity that can drive rip currents. Such problems have been studied theoretically by Liu and Mei (1976) and Dalrymple *et al.* (1977), and theoretically and experimentally by Wind and Vreugdenhil (1986).

The second class of models involves wave-wave interactions. Bowen (1969b) found that the nonlinear interaction between edge waves and normally incident waves produces an a longshore variation of the wave height which, in turn, generates rip currents. This was verified in a laboratory experiment by Bowen and Inman (1969). Similarly, Dalrymple (1975) showed that two intersecting wave trains lead to a short-crested wave pattern which leads to rip currents. Experimental evidence of this driving mechanism was given by Dalrymple and Lanan (1976) and Hammack et al. (1991).

The third class of models hypothesizes that the state of longshore uniform wave set-up could be unstable under certain conditions and that the instability manifests itself in terms of periodic nearshore circulation cells. Early attempts at showing that steady horizontal circulation cells could be produced by perturbing the basic state of a longshore uniform set-up were made by LeBlond and Tang (1974), Hino (1974), and Iwata (1976). Miller and Barcilon (1978) and Dalrymple and Lozano (1978) questioned some of the assumptions made in earlier works and refined the models. Since these models solve eigenvalue problems, they only predict the spacing between the rips. They do not predict the strength of the rip currents. Dalrymple and Lozano included the refraction of the waves by the rip currents which provides a physical mechanism for the maintenance of the circulation cells. They also showed that their predicted rip current spacing compared favorably with field data. As pointed out by Dalrymple and Lozano, these models do not address the initiation mechanisms for the circulation patterns. They only show that circulation patterns other than that of an a longshore uniform set-up can exist on a beach.

5.4. *Models with vertical flow resolution*

Undertow

On a long, straight coast with no variation in the longshore (y-) direction, the net cross-shore mass flux $\overline{Q}_x = 0$. Nevertheless, there is a cross-shore circulation associated with the mass flux Q_{wx} near the surface. This circulation is particularly strong in the surf-zone where Q_{wx} is enhanced by the breaking process. Figure 11 shows the principal flow pattern with a strong undertow (generally of the order 8–10 % of $\sqrt{gh}$) near the bottom to compensate for the shoreward mass flux of the waves.

This circulation pattern was first described quantatively by Dyhr-Nielsen and Sorensen (1970) for the special case of a barred profile, and analyzed

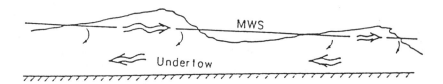

Fig. 11. The circulation flow in the vertical plane including the undertow. (Svendsen, 1984b)

theoretically for a plane beach by Svendsen (1984b). Addtional contributions to the clarification of the phenomenon have been provided by Dally and Dean (1984, 1986), Hansen and Svendsen (1984), Stive and Wind (1986), Svendsen et al. (1987), Okayasu et al. (1988), and Deigaard and Fredsøe (1989), to mention few.

It has been found that the undertow is a balance among the forces on the fluid particle caused by a combination of the radiation stress, the pressure gradient from the sloping mean water surface, and the turbulent shear stresses. Over most of the water column, the turbulent stresses are dominated by the breaker-generated turbulence. Near the bottom, however, the turbulence intensity is small and dominated by the bottom boundary layer (Svendsen et al., 1987; Okayasu et al., 1988). The boundary layer also generates a steady streaming which is of no significance in the surf zone but is important outside the surf-zone (Putrevu and Svendsen, 1993). Finally, the effect of the disturbance of the wave motion due to the dacay of the wave height was addressed by Deigaard and Fredsøe (1989). In essence, they studied the modification of a sine wave motion required to transport wave energy to the region of energy dissipation, particularly the surface roller. The modification due to the variation in depth was discussed by Putrevu and Svendsen (1993).

Basically, the equation solved is obtained from Eq. (43) by assuming steady, 2D vertical flow:

$$-\frac{\partial}{\partial z}\left(\overline{u'w'}\right) = g\frac{d\bar{\zeta}}{dx} + \frac{\overline{\partial u_w^2}}{\partial x} + \frac{\overline{\partial u_w w_w}}{\partial z} \qquad (70)$$

where $\overline{u'w'}$ represents the Reynolds stresses. Various approximations have been used by different authors for the terms in this equation. Generally, however, the Reynolds stresses are modeled using an eddy viscosity ν_t, which reduces

Eq. (70) to

$$\frac{\partial}{\partial z}\left(\nu_t \frac{\partial U}{\partial z}\right) = g\frac{d\bar{\zeta}}{dx} + \frac{\overline{\partial u_w^2}}{\partial x} + \frac{\overline{\partial u_w w_w}}{\partial z}. \tag{71}$$

In most models, the variation of ν_t is parameterized from comparison with measurement. Deigaard et al. (1991) though used a one-equation turbulence closure to assess the variation of the eddy viscosity.

Also, the proper boundary conditions have been discussed. Equation (71) is a second-order equation in the z-direction and hence requires two boundary conditions. The first of those is a condition for the velocity at the bottom, assuming the bottom shear stress is linked to the bottom velocity by a friction factor f (see Sec. 4) which, assuming a small boundary layer thickness and sinusoidal short-wave motion, gives the condition

$$\left.\nu_t \frac{\partial U}{\partial z}\right)_{z=-h_0} \approx \frac{\tau_b}{\rho} = \frac{2}{\pi}fu_{wb}U_b. \tag{72}$$

An alternative to Eq. (72) is to specify a variation of ν_t at the bottom that is compatible with the variation in a bottom boundary layer including $\nu_t \to 0$ at the bottom. If so, the condition $U_b = 0$ is the appropriate bottom boundary condition.

Being short-wave-averaged, the equation only applies to the region below trough level (or mean water level — see the discussion in Sec. 2). The second boundary condition can either be a shear stress at the trough level (mean water level), representing the contribution of the radiation stress and pressure gradient above that level, or it can be the requirement that

$$\int_{-h_0}^{\bar{\zeta}} U dz = -Q_w. \tag{73}$$

Equation (71) has the closed form solution

$$U(x,z) = \int_{-h_0}^{z} \frac{1}{\nu_{tz}} \int_{-h_0}^{z} \alpha_1 dz dz + C_1 \int_{-h_0}^{z} \frac{1}{\nu_{tz}} dz + U_b \tag{74}$$

where C_1 and U_b are integration constants, and α_1 represents the right-hand side of Eq. (71). Figure 12 shows a comparison between Eq. (74) and measurements for a case where a linear variation over depth is assumed for ν_t (Okayasu et al., 1988). It is important to emphasize that with such a variation of ν_t the

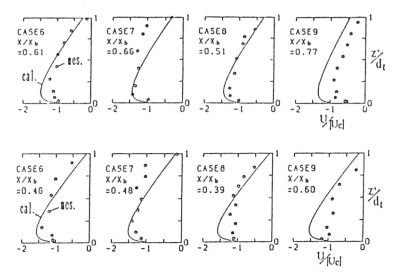

Fig. 12. Comparison between measured and predicted undertow profiles on a 1/30 slope. (Okayasu et al., 1988)

condition $U_b = 0$ can be satisfied and Eq. (74) describes the mean flow inside the bottom boundary layer as well.

A more complete solution of Eq. (43) was given by Svendsen and Lorenz (1989) who also analyzed the longshore current velocity profiles and hence established that the total current profiles essentially have a 3D structure. They also found that for the longshore currents, the variation over depth is much weaker relative to the total velocity in that direction than that for the undertow. Due to the entirely different conditions (strong breaker-generated turbulence over most of the profile combined with the weak turbulence in the bottom boundary layer), the longshore current velocity profile is also distinctly different from the logarithmic velocity profile of open channel flow.

Quasi-3D Models

Models for short-wave-averaged flow with resolution of the vertical current structure have been presented by DeVriend and Stive (1987), Sanchez-Arcilla et al. (1990), Svendsen and Putrevu (1990), Sanchez-Arcilla et al. (1992), Van Dongeren et al. (1994), among others.

Though far from similar in approach, these models are all based on representing the vertical structure of the current velocities and their direction through some solution to the chosen approximation of the local wave-averaged Eq. (43). This structure is then used in the solution for the depth-integrated, short-wave averaged equations. With the exception of Van Dongeren *et al.*, however, all the quasi-3D works assume that the interaction between currents and between currents and waves can be neglected, not only in the determination of the vertical current structure (which is probably a good approximation), but also in the integration of the 2D-horizontal equations (which is probably a less realistic assumption). Figure 13 shows as an example of the variation of the 3D current velocity profiles in the time-varying flow initiated by the onset of longshore currents on a long beach.

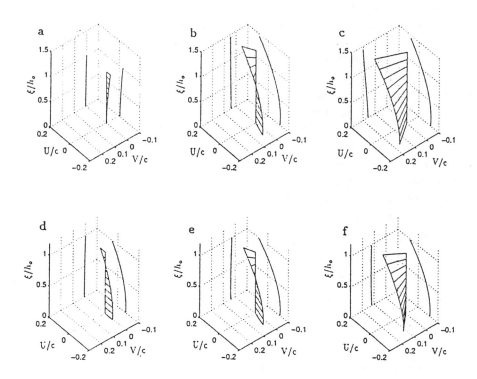

Fig. 13. Three-dimensional current velocity profiles at three different times during the start-up of a longshore current on a long straight beach. (a)–(c) at 25% of breaker depth, (d)–(f) near the breaker line. (Van Dongeren *et al.*, 1994).

These models potentially describe so many of the nearshore circulation phenomena that substantial numerical experiments should be carried out to explore their relevance and accuracy. This would also encompass a thorough validation by a comparison with laboratory and field data, which has not been carried out yet. In addition, further development of many of the model components and modification of underlying assumptions are likely to be needed. This particularly applies to the addition of the dispersive mixing mechanism described in Sec. 5.2.

6. Infragravity Waves

6.1. *Introduction*

The class of gravity waves with periods ranging from about 20 to 200 seconds have come to be known as infragravity waves. The first observations of infragravity motions were reported by Munk (1949b) and Tucker (1950) who coined the term "surf beats" to denote these motions. Later field measurements (e.g., Wright *et al.*, 1979, 1982; Huntley *et al.*, 1981; Holman 1981; Guza and Thornton, 1982, 1985; Oltman-Shay and Guza, 1987; Howd *et al.*, 1991) have clearly revealed that infragravity motions are ubiquitous in the nearshore. In the swash zone, the energy at infragravity frequencies often dominates, exceeding the energy at wind-wave frequencies (Wright *et al.*, 1982; Guza and Thornton 1982, 1985).

Infragravity waves are solutions to the depth-integrated, short-wave-averaged equations of continuity and momentum (Eqs. (7) and (13) of Sec. 2). For the linearized versions of these equations, $\bar{Q}$ can be eliminated to yield the following equation for the infragravity surface elevation (Foda and Mei, 1981; Symonds *et al.*, 1982; Schäffer and Svendsen, 1988)

$$\frac{\partial^2 \zeta}{\partial t^2} - \frac{\partial}{\partial x_\alpha}\left(gh_o\left(\frac{\partial \zeta}{\partial x_\alpha}\right)\right) = \frac{1}{\rho}\frac{\partial^2 S_{\alpha\beta}}{\partial x_\alpha \partial x_\beta} \tag{75}$$

where the radiation stress ($S_{\alpha\beta}$) could vary with time if the short-wave height varies with time (and where for convenience we have omitted the overbar on ζ).

Most of the work on infragravity motions assumes that the bathymetry does not vary in the longshore direction. Hence, the discussion below also assumes longshore uniform bottom bathymetry.

The solutions to the homogeneous version of Eq. (75) (frequently referred to as free waves) are the normal modes of oscillation on a beach. These modes are separated into two kinematically distinct classes. The first class — edge waves — have a discrete set of eigenvalues in the range $\sigma^2/g \leq k_y \leq \sigma^2/gh_x$ where σ is the frequency and k_y is the longshore wave number (Ursell, 1952). The second class — leaky waves — have a continuous spectrum of eigenvalues in the range $\sigma^2 > gk_y$.

Edge waves (discussed in Sec. 6.2) are infragravity waves that are trapped in the nearshore region by refraction.[b] Leaky waves, on the other hand, are infragravity motions that escape back into the deep ocean upon reflection from the shoreline.

The solutions to the inhomogeneous version of Eq. (75) represent forced infragravity motions. Offshore of the surf-zone, in intermediate water depth (relative to the infragravity waves), the solution to the inhomogeneous version of Eq. (75) is referred to as bound waves. Bound infragravity waves are locally generated by the short-wave groups and propagated along with the groups (or are "bound" to the groups). The solutions for the bound waves on a horizontal bottom were given by Longuet-Higgins and Stewart (1962, 1964), Hasselmann (1962), and Gallagher (1971) (see also Bowen and Guza (1978)). These solutions will not be discussed here. Inside the surf-zone, mechanisms that cause the RHS of Eq. (75) to be nonzero and hence force infragravity motions have been proposed by Symonds et al. (1982) and Schäffer and Svendsen (1988). These mechanisms are further discussed in Sec. 6.4.

6.2. Edge waves

On a plane beach, the solutions for the free-edge wave modes were given by Stokes (1846), Eckart (1951), and Ursell (1952). Stokes gave the solution for the so-called zero-mode edge wave (the first normal mode of oscillation of Eq. (75)). Eckart gave the solution for all the edge wave modes assuming that the depth is shallow everywhere (relative to edge wave scales). The extension of Eckart's solution to arbitrary depth was given by Ursell who showed that edge waves follow the dispersion relationship

$$\sigma^2 = gk_y \sin\left[(2n+1)h_x\right] \qquad (76)$$

where n is the mode number of the edge waves. He further showed that the

---

[b]The reader is referred to Holman (1984) and Oltman-Shay and Hathaway (1989) for more detailed discussions than that given below of the basic solution for edge waves, and to Schäffer and Jonsson (1992) for a discussion of edge wave solutions from a geometric optics standpoint.

wave numbers of the edge waves are constrained to the range

$$\frac{\sigma^2}{g} \leq k_y \leq \frac{\sigma^2}{gh_x}. \tag{77}$$

A consequence of this constraint is that no edge waves exist with wave numbers higher than (wave lengths smaller than) those given by the mode-zero edge wave dispersion relationship.

Experimental and Theoretical Analysis of Edge Wave Properties

Some of the interest in edge waves stems from the fact that they have been postulated to be responsible for many features observed on natural beaches. Examples include rip currents (Bowen and Inman, 1969) and observed topographic features (e.g., Bowen and Inman, 1971; Guza and Inman, 1975; Bowen, 1980; Guza and Bowen, 1981; Holman and Bowen, 1982).

Motivated by the possible consequences of the presence of edge waves, much effort has been spent since the 1970s, trying first to prove the existence (or the lack thereof) of edge waves in a natural surf-zone, later to unveil their dominant properties. Since edge waves are very difficult to generate in the laboratory in a controlled way, much of the effort has been oriented towards field data and towards theoretical modeling. While analysis of measurements from cross-shore arrays showed substantial amounts of infragravity energy in the surf-zone, the presence of edge waves could not be conclusively inferred because close to the shore the cross-shore structure of the edge wave modes and leaky waves modes are very similar (see Holman (1983) for a discussion).

By analyzing the longshore structure of the infragravity motions, Huntley *et al.* (1981) provided the first compelling evidence for the presence of edge waves in a natural surf-zone. They concentrated on a subset of the data collected during the NSTS experiments and demonstrated that edge waves could be detected in the measurements of the longshore currents. Based on their analysis, Huntley *et al.*, concluded that "only one progressive edge wave dominates at any particular frequency". Oltman-Shay and Guza (1987) demonstrated, using synthetic data testing, that at any frequency longshore velocity variance would indeed be dominated by one edge wave mode even if all edge wave modes were present.

An extensive analysis by Oltman-Shay and Guza of the data collected during the NSTS experiments showed that edge waves were always present during these experiments. Oltman-Shay and Guza found that the longshore current variance was dominated by low mode edge waves while the cross-shore current was dominated either by higher mode (≥ 3) edge waves or by leaky waves (because of the limitations of the data set, they were not able to distinguish between high-mode edge waves and leaky waves). Their analysis suggested that all the low-mode edge waves (≤ 2) had comparable amplitudes at the shoreline. Combining this with the result that the longshore velocity variance would be dominated by one edge wave mode, they showed that if all the edge wave modes have comparable shoreline amplitudes (a white spectrum), the longshore current at a particular location will be dominated by the lowest edge wave mode that is not trapped significantly shoreward of that location.

Thus, the work of Huntley et al., and Oltman-Shay and Guza demonstrated the presence of low-mode edge waves in the surf-zone. Recent work by Elgar et al. (1992), Herbers et al. (1992), and Oltman-Shay (1994) has provided evidence that high-mode edge waves are also present in the surf-zone.

Recent work has also clarified the effects that nonplanar bottom topography and longshore currents have on edge waves. The effect of the topography was studied by Holman and Bowen (1979), and Kirby et al. (1981) while Howd et al. (1992) and Falques and Iranzo (1992) studied the effects of the longshore current. Oltman-Shay and Howd (1993) included both these effects in a model-data comparison.

It was found that the concave beach face that often is found in nature substantially influences both the dispersion relationship and the cross-shore structure of the edge waves (Holman and Bowen 1979, Oltman-Shay and Howd 1993). In particular, Holman and Bowen showed that (1) assuming a simple beach profile could result in errors up to 100% for the wave number and (2) the relative importance of the longshore and cross-shore velocities is altered by the beach concavity. Oltman-Shay and Howd showed that the influence of concave beach faces on the cross-shore structure of the edge waves occurs even at frequencies where the wave numbers are not substantially altered by the topography. An example calculation showed that assuming a simple beach profile could overestimate the shoreline cross-shore velocity variance by a factor of four (see their Fig. 5).

These results are similar to the findings of Kirby et al. (1981) who investigated the behavior of edge waves in the presence of longshore sand bars. Their results indicated that, for the topographies they considered, the edge wave dispersion relationship was unaltered from the plane beach case. They suggested that, on barred beaches, the edge wave dispersion relationship is controlled by the mean beach slope (as opposed to the class of beach profiles investigated by Holman and Bowen (1979)). Their calculations also demonstrated that even though the dispersion relationship is unchanged, the cross-shore structure of the edge waves is considerably altered by the presence of sand bars. They found that the edge wave surface elevation (cross-shore velocity) profile adjusts itself so that the antinodes (nodes) occur at the bar locations.

Restricting themselves to longshore currents that have $V_{max}/c < 1$, Howd et al. (1992) demonstrated that the cross-shore shapes of the edge waves are sensitive to the presence of a current. They found that for typical field conditions the presence of the longshore current could change the a longshore wave number of the edge waves by up to 30%. This result suggests that, when analyzing field measurements, it is important to account for the presence of the longshore current. An important consequence of the presence of the longshore current is that it introduces an asymmetry: the longshore wave number increases for a longshore current opposing the edge wave propagation and the wave number decreases for a longshore current in the direction of edge wave propagation (see also Oltman-Shay and Guza (1987)). These results were confirmed by Oltman-Shay and Howd (1993) who showed that the effect of the shear in the longshore current is particularly dramatic on the longshore velocity of edge waves. Their results also showed that for the NSTS data sets assuming a simple topography (plane beach) and neglecting longshore currents could lead to errors as high as 45% while estimating the variances at the shoreline.

Falques and Iranzo (1992) removed the restriction on the strength of the longshore current imposed by Howd et al., and demonstrated that, in most cases, the offshore extent of the edge waves increases (decreases) when they propagate against (with) the current. They also demonstrated analytically that in the case of edge waves propagating with a strong longshore current (strong enough so that $V = c$ at some locations), the edge wave is tightly bound to the shore — it cannot extend seaward of the location where $V = c$. Similarly, there is a region in wave number-Froude number space in which edge waves cannot propagate against the current.

Field observations suggest that many quantities characterizing infragravity waves in the surf-zone (wave heights, velocity fluctuations, run-up) are well correlated with the offshore incident wave height (Holman, 1981; Guza and Thornton, 1982, 1985; Holman and Sallanger, 1985; Howd et al. 1991) which points at the short-wave motion as one of the sources of the infragravity waves. Guza and Thornton (1982) showed that the infragravity swash amplitude varied linearly with the incident wave height. Holman and Sallanger analyzed data from a different beach and found that, while the infragravity swash amplitude did depend on the offshore wave height, the variation was more complicated than that found by Guza and Thornton. In particular, they suggested that the normalized (by the offshore significant wave height) infragravity swash amplitude depends on the "Surf Similarity" parameter for the short-wave motion. Similar results were obtained by Guza and Thornton (1985) and Howd et al. (1991)

Recent observations of infragravity energy offshore of the surf-zone (Elgar et al., 1992; Herbers et al., 1992; Oltman-Shay, 1994) may have interesting implications for the importance of surf-zone forcing of infragravity motions. Elgar et al., show that, offshore of the surf-zone, the locally forced bound waves contribute very little to the total infragravity energy except under the rare condition of very energetic swell (see also Okihiro et al. (1992)). This observation indicates that free (or nonlocally forced) infragravity motions are frequently the dominant source of infragravity energy offshore of the surf-zone. Furthermore, Herbers et al., showed that for swell conditions the free infragravity energy at an offshore location is well correlated with and depends linearly on the incident swell energy (see also Oltman-Shay (1994)). Present models of surf-zone generation of infragravity motions (discussed below) suggest that the infragravity energy generated in the surf-zone will exhibit a linear dependence on the incident wave energy. Thus, at present there are strong indications that a substantial amount of infragravity wave generation may be taking place in the surf-zone.

6.3. *Generation of infragravity waves in the surf-zone*

Offshore of the surf-zone, Longuet-Higgins and Stewart (1962, 1964), Hasselmann (1962), Gallagher (1971), and Bowen and Guza (1978) showed that the group structure of the incident wave field could generate infragravity motions. Surf zone generation of one-dimensional infragravity motions (leaky waves) was considered by Symonds et al. (1982) and Schäffer and Svendsen (1988). The first of these works considered the generation of infragravity motions due to temporal variations of the break point on a plane beach due to the variations of the short-wave height. Briefly, Symonds et al., argued that since individual waves in a group have different heights, they are likely to begin breaking at different locations and have different heights at the initial location of breaking. Hence, the position of the break point is a function of time. Symonds et al., assumed that the group structure of the incident waves is destroyed by the breaking process. Thus, the long-wave generation only takes place in a narrow region which is termed as the "zone of initial breaking".

The time variation of the break point generates long waves at the group period (and its higher harmonics) which are radiated both shorewards and seawards. The shoreward-radiated waves are reflected at the shoreline and radiate back out seaward. The model of Symonds et al., predicts that the amplitude of this outgoing free wave is a strong function of the frequency of the long wave. On the other hand, the shoreline amplitude of the infragravity motion was found not to exhibit frequency dependence. They attribute this lack of resonant behavior to the radiation of energy seaward. Symonds and Bowen (1984) extended the model to include the presence of an a longshore bar.

The following heuristic argument is often used to explain the mechanism proposed by Symonds et al. (1982). As discussed previously, the loss of the cross-shore-directed momentum flux due to the breaking process is balanced by changes in the mean water level (set-up). Bowen et al. (1968) showed that, under simplified circumstances, this set-up is proportional to the breaking wave height. Thus, temporal variations of the breaker height lead to corresponding temporal variations of the set-up. This time-varying set-up is equivalent to the infragravity wave-surface elevation.

Looking at the other extreme situation relative to Symonds et al., Schäffer and Svendsen (1988) assumed that all the waves in a group begin to break at the same fixed location. This implies that all the groupiness of the incident wave field is transmitted into the surf-zone. Hence, they could study the generation of infragravity waves due to wave-height variation throughout the surf-zone, and in parallel with Symonds et al., they found that the groupiness of the broken-wave field generates long waves at the group period. This generation can be very strong, and they quantified the strength of the generation in terms of a "reflection coefficient", defined as the ratio between the outgoing free long wave and the incoming bound long wave.

In Schäffer and Svendsen's model, a bound long wave generated by the Longuet-Higgins and Stewart mechanism is assumed incident at the seaward boundary of the model domain. Further generation of the long waves takes place as the incident short waves shoal over the sloping bottom after they break. Thus, the "reflection coefficient" could be more descriptively termed as the "amplification factor" for the nearshore region because it measures how much the incident (set-down) waves are amplified by the nearshore processes of shoaling and breaking. They find that the "reflection coefficient" can attain values as high as 25–30 suggesting very strong generation of infragravity wave energy in the nearshore region. Though some of this amplification is due to the long-wave generation on the slope, Schäffer and Svendsen found that a large fraction of the outgoing free long wave is indeed generated in the surf zone.

Schäffer (1993) combined the models of Symonds *et al.* (1982) and Schäffer and Svendsen to allow for both a time variation of the break point as well as a partial transmission of the groupiness into the surf-zone. He compared the predictions of this model with laboratory measurements of Kostense (1984) and found qualitative agreement. He also found that his model consistently overpredicts the long-wave generation. Schäffer attributes this overprediction to his neglect of bottom friction and the feedback between the long waves and the short waves.

An extension of this model to two horizontal dimensions (Schäffer, 1994) shows that for small angles of incidence leaky waves are generated whereas, for larger angles of incidence edge waves are generated by the same mechanisms. Schäffer (1994) compared his predictions of the amplitude of the forced edge wave with the laboratory measurements of the same quantity by Bowen and Guza (1978). Schäffer's solution assumes a beach of infinite a longshore extent so that the edge waves generated are in a steady state, whereas the experiments conducted by Bowen and Guza had limited longshore extent which prevented the development of steady edge waves. This is probably the reason why Schäffer predicts much higher edge wave amplitudes than the laboratory measurements of Bowen and Guza.

Definitive field evidence for the importance (or the lack thereof) of the generating mechanisms discussed above has not yet been presented. On the one hand, List (1992) solved Eqs. (7) and (14) numerically assuming that the groupiness of the short-wave field is destroyed by the breaking process (thus, his calculations do not include the Schäffer and Svendsen mechanism). He compared his numerical solutions with field data and found that, for the particular data set he was working with, the long wave generated by the Symonds *et al.*, mechanism is secondary to the bound long waves. This contradicts the results discussed in Sec. 6.2 of Herbers *et al.* (1992) and Oltman-Shay (1994), who suggested that the linear correlation between the variances of the infragravity and swell waves indicates that the free infragravity wave variance is predominantly generated in the nearshore. As List pointed out, however, his conclusions are only valid for the conditions of the data set that he analyzed. For example, his numerical experiments show that the importance of the long wave forced by the break point variations increases with increasing beach slope.

It is also pointed out here that the theories of Symonds et al., and Schäffer and Svendsen assume that the forcing wave groups are steady. This will usually not be true in practice. Thus, these theories have to be extended to unsteady wave groups before their importance in the field can be determined.

6.4. *Velocity profiles in infragravity waves*

Putrevu and Svendsen (1995) and Svendsen and Putrevu (1994b) presented a local solution for the vertical structure of the velocity profiles in infragravity waves. Their solution predicts that if there is no local forcing of infragravity waves, then the infragravity velocity profiles will not exhibit vertical structure. On the other hand, the solution predicts that the infragravity waves will exhibit vertical structure if there is local forcing of infragravity waves by short-wave variations. Thus, the Putrevu and Svendsen solution predicts vertical structure in the infragravity velocity field in the zone of initial breaking as well as throughout the surf-zone if the groupiness of the incident short-wave field is not completely destroyed by the breaking process.

7. Shear Waves

Analysis of data collected during the SUPERDUCK field experiment (Crowson et al., 1988) by Oltman-Shay et al. (1989) demonstrated the presence of low frequency wave like oscillations of the longshore currents. These oscillations were found to be progressively longshore with longshore wave lengths and periods of 100 meters and 100 seconds, respectively. For example, Fig. 14 shows a typical longshore wavenumber-frequency spectrum of the longshore velocity (see Oltman-Shay et al., (1989) for details). The solid lines in this figure represent the dispersion relationships of the various edge wave modes and the rectangular boxes show the locations and half power bandwidth of the estimated variance peaks.[c]

As discussed in Sec. 6.2, no surface gravity motions exist for wave numbers

---

[c]Though shear waves were first detected in the data collected during the SUPERDUCK field experiment, there is evidence that these motions were also present during the NSTS experiments (Dodd et al., 1992).

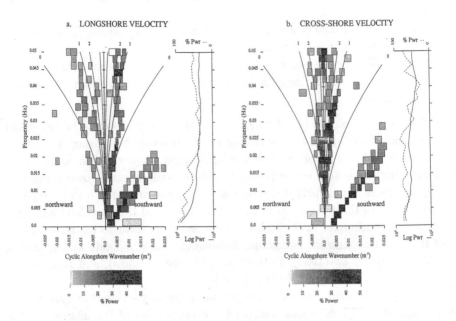

Fig. 14. An estimated frequency-a longshore wave number spectrum for low frequency motions in the surf-zone. (Oltman-Shay et al., 1989)

higher than (or wave lengths lower than) those given by the mode-0 edge wave relationship. The field data in Fig. 14 clearly shows the presence of motions with wave numbers too short to be surface gravity motions (the linear dispersion line to the right of the mode-0 edge wave curve in Fig. 14). Thus, Oltman-Shay et al., concluded that the observed concentration of variance located outside the mode-0 edge wave curve is derived from a source other than surface gravity motions. The motions represented by these dispersion lines have come to be known as shear waves. Observations have shown that shear waves can be quite energetic with velocity amplitudes greater than 30 cm/s (Oltman-Shay et al.) and can contribute up to 33% of the longshore current variance (Howd et al., 1991).

Oltman-Shay et al. demonstrated that shear waves have the following characteristics:

1. They are almost nondispersive and are propagated in the direction of the longshore current.

2. Their kinematics are closely linked to the strength of the mean longshore current.
3. The speed of propagation is in the range $0.5V_p$ - V_p where V_p is the peak longshore current magnitude.
4. Longshore and cross-shore velocity components are in the quadrature phase.
5. The magnitude of the longshore current does not affect the range of observed wave numbers.

Bowen and Holman (1989) suggested that shear waves may be generated by a shear instability of the mean longshore current profile. Using very simple variations of the longshore current and bottom topography, Bowen and Holman showed that longshore currents could be potentially unstable over a range of wave numbers. They further demonstrated that the predicted scales compared favorably with those observed in the field. Bowen and Holman also found that the critical parameter in the stability problem is the shear to the seaward side of the longshore current maximum. Subsequent work by Dodd and Thornton (1990), Dodd et al. (1992), Putrevu and Svendsen (1992a), and Falques and Iranzo (1994) have confirmed these results.

Dodd et al., solved the stability equation using measured longshore current and depth profiles and compared the predictions of the instability theory (frequencies and wave numbers) to measurements made during NSTS and SUPERDUCK experiments. They found that the predictions of the instability theory compared well with the observations at SUPERDUCK. With the NSTS data set, however, Dodd et al., found that the instability theory predicted no generation of shear waves whereas field data clearly showed the presence of shear waves.

Putrevu and Svendsen (1992a) studied the effect of bottom topography on the stability characteristics of longshore currents. Their results demonstrated that the bottom topography has a considerable influence on the stability characteristics. For example, their results showed that the presence of a bar significantly destabilizes the longshore current. This, in turn, suggests that shear waves are more likely to be observed on barred beaches rather than on plane slopes (a conjecture already made by Bowen and Holman).

Falques and Iranzo (1994) presented an efficient numerical solution to the stability problem which relaxes the rigid lid assumption made in the theoretical works mentioned above. They also included the effects of a horizontal eddy viscosity (neglected in all previous works) and bottom friction (also included by Dodd et al.) using very simple formulations. Their results confirm the strong influence the bottom topography has on the stability characteristics. They also demonstrate that the rigid lid assumption is reasonable as long as the maximum local Froude number (the maximum value of $V/\sqrt{gh}$) is much smaller than unity. (This requirement is met in most cases.) They also found that both the bottom friction and horizontal eddy viscosity can substantially influence the stability characteristics of longshore currents.

Recently, Reniers et al. (1994) presented laboratory experiments on the generation of shear waves. They generated uniform longshore currents using a technique of recirculation similar to that used by Visser (1984). Preliminary results indicated the presence of shear waves on barred beaches. Whereas on a plane slope, they find that shear waves do not develop. Given the close connection between this finding and the theoretical predictions, it seems likely that the generating mechanism of shear waves may indeed be the instability mechanism proposed by Bowen and Holman.

The theoretical studies described above consider the linear stability problem only. Because of the exponential growth predicted by the linear theory, all of the above-mentioned analyzes will only be applicable at the initial stage when the assumption of linearity is applicable. While it may be reasonable to expect that the scales predicted by the linear theory are also applicable later in the development process, a nonlinear analysis is required to predict the amplitude and the spatial structure. Preliminary reports of such nonlinear analyzes are just now appearing in the literature (Allen et al., 1994; Ozkan and Kirby, 1994).

Finally, it should also be mentioned that alternative mechanisms for the generation of very low frequency motions have been proposed by Fowler and Dalrymple (1990) and Shemer et al. (1991). The mechanism of Fowler and Dalrymple essentially consists of the interaction of two wave trains of slightly different frequencies. They show that such an interaction produces a migrating rip current which has a low frequency signature in the wave number-frequency space that is similar to that of the shear waves. Similarly, Shemer et al. show that a side band instability of the incident gravity waves could lead to a modulation of the longshore component of the radiation stress which, in turn, causes a modulation of the longshore current that resembles the observed motions.

8. Comprehensive Quasi-3D Models

Comprehensive quasi-3D nearshore models can be developed on the basis of the equations presented above though this has not yet been fully accomplished. The basis for such models is the following.

A model that solves the depth-integrated, short-wave-averaged equations would represent a nearshore circulation model that would be capable of predicting wave-generated current motions under general topographical and wave conditions. If such a model is combined with a state-of-the-art model that predicts short-wave transformations over arbitrary bottom bathymetry, the resulting model system would represent a comprehensive nearshore model that will be capable of predicting waves and wave-averaged motions in the nearshore. Though significant strides are being made these years, such general model systems still remain to be developed (see Van Dongeren et al. (1994) for a preliminary report on a comprehensive circulation model). Such models will be extensions of the models discussed under "Quasi-3D models" in Sec. 5, and they will consist of the following elements:

1. A component that solves the depth-integrated, short-wave-averaged equations of horizontal momentum giving the 2D-horizontal variation of the current/infragravity-wave pattern. Early versions of such model components were developed by Noda (1974), Noda et al. (1974), Ebersole and Dalrymple (1980), Kirby and Dalrymple (1982), Wu and Liu (1985), and Winer (1988).
2. A component that evaluates the analytical solutions to the vertical distribution of horizontal velocities of the time-varying currents — which are equivalent to the particle velocities of the infragravity waves — and calculates the quantities required for the 2D-horizontal equations under (1) above (for example, the lateral dispersive mixing coefficient).
3. A short-wave transformation model ("wave driver") that describes the propagation of the short-waves in the computational region with particular emphasis on predicting the radiation stresses, mass fluxes, etc., of the short period wave motion. This element essentially determines the short-wave forcing for the equations in (1) and (2).

There are several important features of the 2D-horizontal equations which are worth mentioning.

First of all, virtually no approximations have been made in the derivation of these equations other than the approximations already embedded in the fundamental Reynolds equations for turbulent fluid motion. Thus, they automatically satisfy the (exact) nonlinear boundary conditions at the bottom and the free surface, which means that the effects of bottom friction, wind stresses on the surface, etc., can be incorporated exactly to the extent we need for the sediment processes. This means that we can expect the solutions to these equations to include the actual processes that occur on a real beach even if we do not presently have simple representations (or even knowledge) of these processes.

Secondly, it is now possible to solve these equations using state-of-the-art numerical techniques, which are both highly efficient and accurate, without omitting any of the terms in the equations. This means that we can expect such solutions to include the actual processes that occur on a real beach even if we do not presently have simple representations or even knowledge about these processes. An example of this is the discovery of the importance that the nonlinear current-current and wave-current interactions have for the lateral mixing (Putrevu and Svendsen, 1992b, Svendsen and Putrevu, 1994a) described in Sec. 5.3, and the original discovery of the radiation stress as a term in the equations is another example.

The opportunities this accuracy provides have not been fully explored yet in the literature.

The second integral part of such a model system is the solution for the vertical structure of the currents and infragravity wave particle velocities. As described elsewhere in this paper, it has been found that in the nearshore region currents not only vary in strength over depth (as one would expect) but often also have widely different directions at the bottom and at the surface. This variation also changes quite dramatically over time. An example was shown in Fig. 13.

The lateral dispersive mixing mechanism crucially depends on this vertical variation. However, in addition to influencing the horizontal distribution of the currents (as the mixing does), the vertical variation clearly will also have a strong effect on how currents and long waves move sediments. This is because the major part of the sediment transport occurs close to the bottom, whereas the water transport involved in the large scale current and infragravity-wave patterns is determined by the flow over the entire depth. Hence, the sediment motion generated by the currents and infragravity waves will in general be in a different direction than what is perceived as the current direction if only the depth-averaged motion is considered. Therefore, the resolution of the vertical current structure is crucial not only for the correct modeling of the hydrodynamics but also the sediment transport processes. Hence, if a sediment-transport component is added to such a quasi-3D model we can expect that the model will include these mechanisms.

The third component of the model system is the so-called wave driver. The wave driver can, in principle, be any short-wave transformation model. Though there are exceptions (see, e.g., Watanabe, 1982; Winer, 1988), most nearshore models published so far have assumed a long, straight coast. The wave driver is then replaced simply by Snell's law, which provides the wave pattern, plus the wave-averaged energy equation, which provides the wave-height variation. This approach is not possible on a general bathymetry and for general wave input. A bona fide wave propagation model is needed as a driver for such cases.

Since the short-wave time scale is much smaller than the time scale of the nearshore circulation, the short-wave models suitable as drivers must be reasonably crude in order to be computationally economical, and they are not accurate enough to actually predict the nearshore circulation. They must, however, include a reasonably accurate description of wave-height decay in the surf-zone. Their primary function here is to provide the driving forces for the circulation in the form of radiation stress, mass flux, etc. With this as the forcing, the nearshore circulation can then be determined by solution of the equations described under (1) and (2). However, if a sediment-transport component is added to the quasi-3D model, the wave driver would have the additional function of providing sufficient information about the local wave particle velocity near the bottom to make it possible to determine the net effect of the oscillatory part of the sediment motion.

9. Concluding Remarks

In this review, we have attempted to discuss the present state of understanding of surf-zone hydrodynamics. As it has hopefully become clear from this review, considerable progress has been made in the last thirty years even though there are still many areas where our understanding is far from satisfactory.

Wave breaking provides the forcing for larger scale motions in the surf-zone. It is therefore probably both somewhat ironic as well as unfortunate that at the present time there exists no satisfactory theory to describe breaking and broken waves in the surf-zone. This is currently a topic of intense research interest and we are confident that substantial progress will be made in the near future.

While our overall understanding of wave-induced nearshore circulations seems to be fairly sound, there are a number of phenomena that clearly require further study. These include, but are not limited to, quantitatively accurate predictions of rip currents, the predictions of longshore currents on barred beaches, and the importance of longshore inhomogenieties on nearshore circulations. Once again, these topics are currently being pursued by a number of investigators, and we expect considerable progress in the near future.

Recent work has demonstrated that the surf-zone is an important region for the generation of infragravity motions. While the present indications are that a substantial fraction of the infragravity energy seems to be generated in and near the surf-zone, the existing models of surf-zone generation of infragravity motions have not been verified.

Shear waves seem to be amenable to an interpretation as a manifestation of an instability of the longshore current. Ongoing work on the nonlinear development of the instability and the importance of wave group forcing on these motions promises to yield interesting results.

In conclusion, the subject of surf-zone hydrodynamics is at an exciting stage of development right now and we expect that many of the issues will be clarified in the near future. We may also expect that the ongoing and future work will discover phenomena which we are currently unaware of.

Acknowledgements

We wish to thank N. Kobayashi, J. Oltman-Shay, and D. H. Peregrine for useful and interesting discussions. We also thank K. Verna for editing this manuscript. This effort was supported by the Army Research Office under contract number DAAL03-92-G-0016 (IAS), and the Office of Naval Research, Coastal Sciences under contract number N00014-94-C-0004 (UP). The US government is authorized to produce and distribute reprints for government purposes notwithstanding any copyright notation that may appear herein.

References

Allen, J. S., P. A. Newberger, and R. A. Holman (1994). Nonlinear shear waves in the surf-zone over plane beaches. Abstract O12F-7, *EOS Transactions* **75**: 322.

Allender, J. H., J. D. Ditmars, W. Harrison, and R. A. Paddock (1978). Comparison of model and observed nearshore circulation. *16th Int. Conf. Coastal Engineering*. ASCE. 810–827.

Allender, J. H. and J. D. Ditmars (1981). Field measurements of longshore currents on a barred beach. *Coastal Engrg.* **5**: 295–309.

Banner, M. L. (1987). Surging characteristics of spilling zones of quasi-steady breaking water waves. *IUTAM Symp. Nonlinear Water Waves*, Tokyo, Japan.

Barnes, T. C. D., D. H. Peregrine, and G. Watson (1994). Low frequency wave generation by a single wave group. *Proc. Int. Symp.: Waves — Physical and Numerical Modeling*, Vancouver, Canada. 280–286.

Basco, D. R. and T. Yamashita (1986). Towards a simple model of the wave breaking transition region in surf-zones. *Proc. 20th Int. Conf. Coastal Engineering.* 955.

Battjes, J. A. (1974). Surf similarity. *Proc. 14th Coastal Engineering Conf.* 466–480.

Battjes, J. A. (1975). Modelling of turbulence in the surf-zone. *Proc. Symp. Modelling Techniques*, San Francisco, USA. ASCE. 1050–61.

Battjes, J. A. (1988). Surf-zone dynamics. *Ann. Rev. Fluid Mech.* **20**: 257–293.

Battjes, J. A. and J. P. F. M. Janssen (1978). Energy loss and set-up due to breaking of random waves. *Proc. 16th ICCE*, Hamburg, Germany. 569–587.

Battjes, J. A. and T. Sakai (1981). Velocity field in a steady breaker. *J. Fluid Mech.* **111**: 121–137.

Battjes, J. A., R. J Sobey, and M. J. F. Stive (1990). Nearshore circulation. In *The Sea: Ocean Engineering Science*, Vol. 9. John Wiley and Sons. 468–493.

Battjes, J. A. and M. J. F. Stive (1985). Calibration and verification of a dissipation model for random breaking waves. *J. Geophy. Res.* **90** (C5): 9159–9167.

Bowen, A. J. (1969a). The generation of longshore currents on a plane beach. *J. Marine Res.* **27** (2): 206–215.

Bowen, A. J. (1969b). Rip currents, Part 1: Theoretical investigations. *J. Geophys. Res.* **74** (23): 5467–78.

Bowen, A. J. (1980). Simple models of nearshore sedimentation; beach profiles and longshore bars. *The Coastline of Canada*, ed. S. B. McCann, Geological Survey of Canada. Publication 80–10.

Bowen, A. J. and R. T. Guza (1978). Edge waves and surf beat. *J. Geophys. Res.* **83** (C4): 1913–1920.

Bowen, A. J. and R. A. Holman (1989). Shear instabilities in the mean longshore current 1: Theory. *J. Geophys. Res.* **94** (C12): 18,023–18,030.

Bowen, A. J. and D. L. Inman (1969). Rip currents. Part 2: Laboratory and field experiments. *J. Geophys. Res.* **74** (23): 5479–5490.

Bowen, A. J. and D. L. Inman (1971). Edge waves and crescentic bars. *J. Geophys. Res.* **76** (36): 8662.

Bowen, A. J. and D. L. Inman (1974). Nearshore mixing due to waves and wave induced currents. *Rapp. P.-v. Reun. Cons. Int.* **167**: 6–12.

Bowen, A. J., D. L. Inman, and V. P. Simmons (1968). Wave "set-down" and set-up. *J. Geophys. Res.* **73** (8): 2569–2577.

Brocchini, M., M. Drago, and L. Ivoenitti (1992). The modeling of short waves in shallow waters. Comparison of numerical models based on Boussinesq and Serre equations. *Proc. 23rd Int. Conf. Coastal Engineering.*

Bruun, P. (1963). Longshore currents and longshore troughs. *J. Geophys. Res.* **68** (4): 1065–1078.

Cox, D. T., N. Kobayashi, and D. L. Kriebel (1994). Numerical model verification using SUPERTANK data in surf and swash zones. *Proc. Coastal Dynamics '94*. Barcelona, Spain. ASCE.

Cox, D. T., N. Kobayashi, and A. Wurjanto (1992). Irregular wave transformation processes in surf and swash zones. *Proc. 23rd Int. Conf. Coastal Engineering*, Venice, Italy. Ch. 10, 156–169.

Christoffersen, J. B. and I. G. Jonsson (1985). Bed friction and dissipation in a combined current and wave motion. *Ocean Eng.* **12** (5): 387–423.

Church, J. C. and E. B. Thornton (1993). Effects of breaking wave induced turbulence within a longshore current model. *Coast. Eng.* **20** (1): 1–28.

Church, J. C., E. B. Thornton, and J. Oltman-Shay (1994). Mixing by shear instabilities of the longshore current. Submitted to *J. Geophys. Res*.

Crowson, R. A., W. A. Birkemeier, H. M. Klein, Harriet, and H. C. Miller (1988). SUPERDUCK Nearshore Processes Experiment: Summary of Studies CERC Field Research Facility. US Army Engineer Waterways Experiment Station, Coastal Engineering Research Center, Vicksburg, Mississippi, USA. Technical Report CERC-88-12.

Dally, W. R. (1990). Random breaking waves: A closed form solution for planar beaches. *Coast. Eng.* **14** (3): 233–263.

Dally, W. R. (1992). Random breaking waves: Field verification of a wave by wave algorithm for engineering application. *Coast. Eng.* **16** (4): 369–397.

Dally, W. R. and R. G. Dean (1984). Suspended sediment transport and beach profile evaluation. *J. Wtrwy., Port, Coast. and Oc. Eng., ASCE* **110** (1): 15–33.

Dally, W. R. and R. G. Dean (1986). Discussion of mass flux and undertow in a surf-zone, ed. I. A. Svendsen. *Coast. Eng.* **10** (3): 289–299.

Dally, W. R., R. G. Dean, and R. A. Dalrymple (1985). Wave height variation across beaches of arbitrary profile. *J. Geophys. Res.* **90** (6): 11.917–11.927.

Dalrymple, R. A. (1975). A mechanism for rip current generation on an open coast. *J. Geophys. Res.* **80** (24): 3485–3487.

Dalrymple, R. A. (1978). Rip currents and their causes. *16th Int. Conf. Coastal Engineering*, Hamburg, Germany. 1414–1427.

Dalrymple, R. A., R. A. Eubanks, and W. A. Birkemeier (1977). Wave-induced circulation in shallow basins. *J. Wtrwy., Port, Coast. and Oc. Eng.* **103** (1): 117–135.

Dalrymple, R. A. and G. A. Lanan (1976). Beach cusps formed by intersecting waves. *Bull. Geol. Soc. Amer.* **87** (1): 57–60.

Dalrymple, R. A. and C. J. Lozano (1978). Wave current interaction models for rip currents. *J. Geophys. Res.* **83** (C12): 6063.

Davis, A. G., R. L. Soulsby, and H. L. King (1988). A numerical model of the combined wave current bottom boundary layer. *J. Geophys. Res.* **93** (C1): 491–508.

deVriend, H. J. and M. J. F. Stive (1987). Quasi-3D modeling of nearshore currents. *Coast. Eng.* **11** (5&6): 565–601.

Deigaard, R. and J. Fredsøe (1989). Shear stress distribution in dissipative water waves. *Coast. Eng.* **13** (4): 357–378.

Deigaard, R., P. Justesen, and J. Fredsøe (1991). Modelling undertow by a one equation turbulence model. *Coast. Eng.* **15** (5&6): 431–458.

Dodd, N., J. Oltman-Shay, and E. B. Thornton (1992). Shear instabilities in the longshore current: A comparison of observation and theory. *J. Physical Oceanography* **22** (1): 61–82.

Dodd, N. and E. B. Thornton (1990). Growth and energetics of shear waves in the nearshore. *J. Geophys. Res.* **95** (9): 16.075–16.083.

Duncan, J. H. (1981). An experimental investigation of wave breaking produced by a towed hydrofoil. *Proc. Royal Society of London*, Series A, **377**: 331–348.

Dyhr-Nielsen, M. and T. Sorensen (1970). Some sand transport phenomena on coasts with bars. *Proc. 12th Coastal Engineering Conf.* 855–66.

Ebersole, B. A. and R. A. Dalrymple (1980). Numerical modeling of nearshore circulation. *Proc. 17th Int. Conf. Coastal Engineering.* ASCE. 2710-2725.

Eckart, C. (1951). Surface waves on water of variable depth. Scripps Institution of Oceanography, LaJolla, CA. Wave Report 100.

Elder, J. W. (1959). The dispersion of marked fluid in turbulent shear flow. *J. Fluid Mech.* **5**: 544–560.

Elgar, S., T. H. C. Herbers, M. Okihiro, J. Oltman-Shay, and R. T. Guza (1992). Observations of infragravity waves. *J. Geophys. Res.* **97** (C10): 15.573–15.577.

Falques, A. and V. Iranzo (1992). Edge waves on a longshore shear flow. *Phys. Fluids* **A4** (10): 2169–2190.

Falques, A. and V. Iranzo (1994). Numerical simulation of vorticity waves in the nearshore. *J. Geophys. Res.* **99** (C1): 825–841.

Fischer, H. B. (1978). On the tensor form of the bulk dispersion coefficient in a bounded skewed shear flow. *J. Geophys. Res.* **83** (C5): 2373–2375.

Fischer, H. B., E. J. List, R. C. Y. Koh, J. Imberger, and N. H. Brooks (1979). *Mixing in Inland and Coastal Waters.* Academic Press Inc. 483.

Foda, M. A. and C. C. Mei (1981). Nonlinear excitation of long trapped waves by a group of short swells. *J. Fluid Mech.* **111**: 319–345.

Fowler, R. E. and R. A. Dalrymple (1990). Wave group forced nearshore circulation. *Proc. 22nd Coastal Engineering Conf.* 729–742.

Fredsøe, J. (1983). The turbulent boundary layer in combined wave-current motion, *J. Hydraulic Eng.* **110** (8): 1103–1120.

Fredsøe, J. and R. Deigaard (1993). *Mechanisms of Coastal Sediment Transport.* World Scientific. 369.

Gallagher, B. (1971). Generation of surf beat by nonlinear wave interactions. *J. Fluid Mech.* **49**: 1–20.

Galvin, C. J. (1968). Breaker type classification on three laboratory beaches.

J. Geophys. Res. **73** (12).

Galvin, C. J. (1972). Wave brewaking in shallow water. In *Waves on Beaches and Resulting Sediment Transport*, ed. R. E. Meyer. 413-456.

George, R., R. E. Flick, and R. T. Guza (1994). Observations of turbulence in the surf-zone. *J. Geophys. Res.* **99** (C1): 801–810.

Grant, W. D. and O. S. Madsen (1979). Combined wave and current interaction with a rough bottom. *J. Geophys. Res.* **84** (C4): 1797–1808.

Greenwood, B. and D.J. Sherman (1986). Longshore current profiles and lateral mixing across the surf-zone of a barred nearshore. *Coast. Eng.* **10** (2): 149–168.

Grilli S. T., I. A. Svendsen, and R. Subramanya (1994). *Breaking Criterion and Characteristics for Solitary Waves on Plane Beaches* (submitted for publication).

Guza, R. T. and A. J. Bowen (1981). On the amplitude of beach cusps. *J. Geophys. Res.* **86** (C5): 4125.

Guza, R. T. and D. L. Inman (1975). Edge waves and beach cusps. *J. Geophys. Res.* **80** (21): 2997.

Guza, R. T. and E. B. Thornton (1982). Swash oscillations on a natural beach. *J. Geophys. Res.* **87** (C1): 483–491.

Guza, R. T. and E.B. Thornton (1985). Observations of surf beat. *J. Geophys. Res.* **90** (C2): 3161–3171.

Hammack, J., N. Scheffner, and H. Segur (1991). A note on the generation and narrowness of periodic rip currents. *J. Geophys. Res.* **96** (C3): 4909–4914.

Hansen, J. B. (1990). Periodic waves in the surf-zone: Analysis of experimental data. *Coast. Eng.* **14** (1): 19–41.

Hansen, J. B. and I. A. Svendsen (1984). A theoretical and experimental study of the undertow. *Proc. 19th Coastal Engineering Conf.*, Houston, USA. 2246–2262.

Harris, T. F. W., J. M. Jordan, W. R. McMurray, C. J. Verwey and F. P. Anderson (1963). Mixing in the surf-zone. *Int. J. Air and Water Pollut.* **7**: 649–667.

Hasselman, K. (1962). On the nonlinear energy transfer in gravity wave spectrum, Part 1, General theory. *J. Fluid Mech.* **12**: 481–500.

Hattori, M. and T. Aono (1985). Experimental study on turbulence structures under spilling breakers. *The Ocean Surface*, Toba and Mitsuyasu, Reidel Publishing Company, Dordrecht. 419–424.

Herbers, T. H. C., S. Elgar, R. T. Guza, and W. C. O'Riley (1992). Infragravity frequency motions on the shelf. *Proc. 23rd Coastal Engineering Conf.*. 846–859.

Hibberd, S. and D. H. Peregrine (1979). Surf and runup on a beach: A uniform bore. *J. Fluid Mech.* **95**: 323–345.

Hino, M. (1974). Theory on the formation of rip currents and cuspidal coast. *Proc. 14th Int. Conf. Coastal Engineering.* 901–919.

Holman, R. A. (1981). Infragravity energy in the surf-zone. *J. Geophys. Res.* **86** (C7): 6442–6450.

Holman, R. A. (1983). Edge waves and the configuration of the shoreline. *The CRC Handbook of Coastal Processes and Erostion.* CRC Press. 21–33.

Holman, R. A. and A. J. Bowen (1979). Edge waves over complex beach profiles. *J. Geophys. Res.* **84** (C10): 6330–6346.

Holman, R. A. and A. J. Bowen (1982). Bars, bumps and holes: Models for generation of complex beach topography. *J. Geophys. Res.* **87** (C1): 457.

Holman, R. A. and A. H. Sallenger (1985). Set-up and swash on a natural beach, *J. Geophys. Res.* **90** (C1): 945–953.

Horikawa, K. and C. T. Kuo (1966). A study of wave transformation inside the surf-zone. *Proc. 10th Coastal Engineering Conf.* 217–233.

Howd, P. A., A. J. Bowen, and R. A. Holman (1992). Edge waves in the presence of strong longshore currents. *J. Geophys. Res.* **97** (C7): 11,357–11,371.

Howd, P. A., J. Oltman-Shay, and R. A. Holman (1991). Wave variance partitioning in the trough of a barred beach. *J. Geophys. Res.* **96** (C7): 12,781–12,795.

Huntley, D. A., R. T. Guza, and E. B. Thornton (1981). Field observations of surf beat, Part I: Progressive edge waves. *J. Geophys. Res.* **86** (C7): 6451–6466.

Hwang, L.-S. and D. Divoky (1970). Breaking wave setup and decay on gentle slopes. *Proc. 12th Int. Conf. Coastal Engineering.* Ch. 23 (69), 377–389.

Inman, D. L., R. J. Tait, and C. E. Nordstrom (1971). Mixing in the surf-zone. *J. Geophys. Res.* **76** (15): 3493–3514.

Irribarren, C. R. and C. Nogales (1949). Protection des Ports II, *Comm. 4, 17th Int. Navigation Congress*, Lisbon, Portugal. 31–80.

Iwata, N. (1976). Rip current spacing. *J. Oceanographical Soc. Jpn.* **32** (1): 1–10.

Janssen, P. C. M. (1986). Laboratory observations of the kinematics in the aerated region of breaking waves. *Coast. Eng.* **9** (5): 453–477.

Jonsson, I. G. (1966). Wave boundary layers and friction factors. *Proc. 10th Coastal Engineering Conf.* 127–148.

Jonsson, I. G. and N. A. Carlsen (1976). Experimental and theoretical investigations in an oscillatory turbulent boundary layer. *J. Hydraulic Res.* **14** (1): 45–60.

Karambas, Th., Y. Krestenitis, and C. Koutitas (1990). A numerical solution of the Boussinesq equations in the inshore zone. *Hydrosoft* **3** (1): 34–37.

Karambas, Th. and C. Koutitas (1992). A breaking wave propagation model based on the Boussinesq equations. *Coast. Eng.* **18** (1): 1–19.

Kirby, J. T. and R. A. Dalrymple (1982). Numerical modeling of the nearshore region. Ocean Engineering Program, Department of Civil Engineering, University of Delaware, Research Report CE-82-24.

Kirby, J. T., R. A. Dalrymple, and P. L.-F. Liu (1981). Modification of edge waves by barred beach topography. *Coast. Eng.* **5** (1): 35–49.

Kobayashi, N., G. S. DeSilva, and K. D. Watson (1989). Wave transformation and swash oscillation on gentle and steep slopes. *J. Geophys. Res.* **94** (C1): 951–966.

Kobayashi, N. and A. Wurjanto (1992). Irregular wave setup and runup on beaches. *J. Wtrwy., Port, Coast. and Oc. Eng.* **118** (4): 368–386.

Kostense, J. K. (1984). Measurements of surf beat and set-down beneath wave groups. *Proc. 19th Coastal Engineering Conf.* 724–740.

Kuriyama, Y. and Y. Ozaki (1993). Longshore current distribution on a bar-trough beach – Field measurements at HORF and numerical model. Rpt. Port Harbour Res. Inst. **32** (3).

Larson, M. and N. C. Kraus (1991). Numerical model of longshore current for bar and trough beaches. *J. Wtrwy., Port, Coast. and Oc. Eng.* ASCE. **117** (4): 326–347.

Le Blond, P. H. and G. L. Tang (1974). On energy coupling between waves and rip currents. *J. Geophys. Res.* **79** (6): 811–816.

Le Méhauté, B. (1962). On non-saturated breakers and wave run-up. *Proc. 8th Coastal Engineering Conf.* 77–92.

Lippmann, T. C. and R. A. Holman (1992). Wave breaking in the trough of a natural sand bar. *Trans. Amer. Geophys. Union* **73**: 256.

Lin, J.-C. and D. Rockwell (1994). Instantaneous structure of a breaking wave. *Phys. Fluids* **6** (9): 2877–2879.

List, J. H. (1992). A model for the generation of two-dimensional surf beat. *J. Geophys. Res.* **97** (C4): 5623–5635.

Liu, P. L.-F. and R. A. Dalrymple (1978). Bottom frictional stresses and longshore currents due to waves with large angles of incidence. *J. Marine Res.* **36** (2): 357–375.

Liu, P. L.-F. and C. C. Mei (1976). Water motions on a beach in the presence of a breakwater. Parts 1 and 2. *J. Geophys. Res.* **81** (18): 3079–3094.

Longuet-Higgins, M. S. (1970). Longshore currents generated by obliquely incident sea waves. Parts 1 and 2. *J. Geophys. Res.* **75** (33): 6778–6789, 6790–6801.

Longuet-Higgins, M. S. (1973). A model of flow separation at a free surface. *J. Fluid Mech.* **57**: 129–148.

Longuet-Higgins, M. S. and R. W. Stewart (1962). Radiation stress and mass transport in gravity waves with application to 'surf-beats'. *J. Fluid Mech.*, **8**: 565–583.

Longuet-Higgins, M. S. and R.W. Stewart (1964). Radiation stress in water waves, a physical discussion with application. *Deep Sea Res.* **11**: 529–563.

Longuet-Higgins, M. S. and J. S. Turner (1974). An entraining plume model of a spilling breaker. *J. Fluid Mech.* **63** (1): 1–20.

Mei, C. C. (1983). *The Applied Dynamics of Ocean Surface Waves*. John Wiley and Sons. 740.

Mei, C. C. and P.L.-F. Liu (1977). Effects of topography on the circulation in and near the surf-zone – linear theory. *Estuarine and Coastal Marine Science* **5** (1): 25–37.

Miller, C. and A. Barcilon (1978). Hydrodynamic instability in the surf-zone as a mechanism for the formation of horizontal gyres. *J. Geophys. Res.* **83** (C8): 4107–4116.

Munk, W. H. (1949a). The solitary wave theory and its application to surf problems. *Annals of the New York Academy of Sciences* **51**: 376–424.

Munk, W. H. (1949b). Surf beats. *Trans. Amer. Geophys. Union* **30**: 849–854.

Munk, W. H. and R. S. Arthur (1952). Wave intensity along a refracted ray in gravity waves. *Nat. Bur. Sand Circ.* **521**.

Nadaoka, K. (1986). A fundamental study on shoaling and velocity field structure of water waves in the nearshore zone. Ph.D. Dissertation, Department of Civil Engineering, Tokyo Institute of Technology, Tokyo, Japan.

Nadaoka, K., M. Hino, and Y. Koyano (1989). Structure of the turbulent flow field under breaking waves in the surf-zone. *J. Fluid Mech.* **204**: 359–387.

Nadaoka, K. and T. Kondoh (1982). Laboratory measurements of velocity field structure in the surf-zone by LDV. *Coast. Eng. Jpn.* **25**: 125–145.

Nielsen, P. (1992). Coastal bottom boundary layers and sediment transport. World Scientific.

Noda, E. K. (1974). Wave induced nearshore circulation. *J. Geophys. Res.* **79** (27): 4097–4106.

Noda, E. K., C. J. Sonu, V. C. Rupert, and J. I. Collins (1974). Nearshore circulations under sea breeze conditions and wave-current interactions in the surf-zone. Tetra Technical Report TC-149-4.

Okayasu, A. (1989). Characteristics of turbulence structure and undertow in the surf-zone. Ph.D. Dissertation, Department of Civil Engineering, University of Tokyo. 119.

Okayasu, A., T. Shibayama, and K. Horikawa (1988). Vertical variation of undertow in the surf-zone. *Proc. 21st Coastal Engineering Conf.* 478–491.

Okihiro, M., R. T. Guza, and R. J. Seymour (1992). Bound infragravity waves. *J. Geophys. Res.* **97** (C7): 11.453–11.469.

Oltman-Shay, J. (1994). Evidence of high mode edge waves. *J. Geophys. Res.* (submitted).

Oltman-Shay, J. and R. T. Guza (1987). Infragravity edge wave observations on two California beaches. *J. Phys. Oceanography* **17** (5): 644–663.

Oltman-Shay, J. and K. Hathaway (1989). Infragravity energy and its implications in nearshore sediment transport dynamics. Waterways Experiment Station, Vicksburg, MS, USA, Report CERC-89-6.

Oltman-Shay, J. and P. A. Howd (1993). Edge waves on nonplanar bathymetry and a longshore currents: A model and data comparison. *J. Geophys. Res.* **98** (2): 2495–2507.

Oltman-Shay, J., P. A. Howd, and W. A. Berkemeier (1989). Shear instabilities of the mean longshore current 2: Field observations. *J. Geophys. Res.* **94** (12): 18.031–18.042.

Ozkan, H. T. and J. T. Kirby (1994). Numerical Study of finite amplitude shear wave instabilities. Abstract O12F-6, *EOS Trans.* **75**: 322.

Packwood, A. and D. H. Peregrine (1980). The propagation of solitary waves and bores over a porous bed. *Coast. Eng.* **3** (3): 221–242.

Packwood, A. (1983). The influence of beacgh porosity on wave uprush and backwash. *Coast. Eng.* **7**: 29–40.

Peregrine, D. H. (1983). Breaking waves on beaches. *Ann. Rev. Fluid Mech.* **15**: 149–178.

Peregrine, D. H. and I. A. Svendsen (1978). Spilling breakers, bores and hydraulic jumps. *Proc. 16th Int. Conf. Coastal Engineering.* 540–550.

Phillips, O. M. (1977). *The Dynamics of the Upper Ocean.* Cambridge University Press. 336.

Putrevu, U., J. Oltman-Shay, and I. A. Svendsen (1995). Effect of a longshore nonuniformities on longshore current predictions. *J. Geophys. Res.* **100** (C8): 16,119–16,130.

Putrevu, U. and I. A. Svendsen (1992a). Shear instability of longshore currents: A numerical study. *J. Geophys. Res.* **97** (C5): 7283–7303.

Putrevu, U. and I. A. Svendsen (1992b). A Mixing Mechanism in the Nearshore Region. *Proc. 23rd Int. Conf. Coastal Engineering.* 2758–2771.

Putrevu, U. and I. A. Svendsen (1993). Vertical structure of the undertow outside the surf-zone. *J. Geophys. Res.* **98** (C12): 22.707–22.716.

Putrevu, U. and I. A. Svendsen (1995). Infragravity velocity profiles in the surf-zone. *J. Geophys. Res.* **100** (C8): 16,131–16,142.

Reniers, A. J. H. M., J. A. Battjes, A. Falques and D. A. Huntley (1994). Shear wave laboratory experiment. *Proc. Int. Symp.: Waves — Physical and Numerical Modeling,* Vancouver, Canada. 356–365.

Sanchez-Arcilla, A., F. Collado, M. Lemos, and F. Rivero (1990). Another quasi-3D model for surf-zone flows. *Proc. 22nd Coastal Engineering Conf.* Ch. 24, 316–329.

Sanchez-Arcilla, A., F. Collado, and A. Rodrigues (1992). Vertically varying velocity field in Q-3D nearshore circulation. *Proc. 23nd Coastal Engineering Conf.* Ch. 215, 2811–2824.

Schäffer, H. A. (1993). Infragravity waves induced by short-wave groups. *J. Fluid Mech.* **247**: 551–588.

Schäffer, H. A. (1994). Edge waves forced by short-wave groups. *J. Fluid Mech.* **259**: 125–148.

Schäffer, H. A. and I. G. Jonsson (1992). Edge waves revisited. *Coast. Eng.* **16** (4): 349–368.

Schäffer, H. A. and I. A. Svendsen (1986). Boundary layer flow under skew waves. Institute of Hydrodynamics and Hydraulic Engineering, Technical University of Denmark, Progress Report 64. 13–23.

Schäffer, H. A. and I. A. Svendsen (1988). Surf beat generation on a mild slope beach. *Proc. 21st Int. Conf. Coastal Engineering.* 1058–1072.

Schäffer, H. A., R. Deigaard, P. A. Madsen (1992). A two-dimensional surf-zone model based on the Boussinesq equations. *Proc. 23rd Int. Conf. Coastal Engineering.*

Schäffer, H. A., P. A. Madsen, and R. Deigaard (1993). A Boussinesq model for waves breaking in shallow water. *Coast. Eng.* **20** (3&4): 185–202.

Shemer, L., N. Dodd and E. B. Thornton (1991). Slow-time modulation of finite depth nonlinear water waves: Relation to longshore current oscillations. *J. Geophys. Res.* **96** (C4): 7105–7113.

Shepard, F. P., K. O. Emery, and E. C. LaFond (1941). Rip currents: A process of geological importance, *J. Geology* **49**: 337–369.

Shepard, F. P. and D. L. Inman (1950). Nearshore circulation. *Proc. 1st Int. Conf. Coastal Engineerng.* 50–59.

Sleath, J. F. A. (1984). Sea bed mechanics. John Wiley and Sons. 335.

Smith, J. M., M. Larson, and N. C. Kraus (1993). Longshore current on a barred beach: Field measurements and calculation. *J. Geophys. Res.* **98** (C12): 22,717–22,731.

Sonu, C. J. (1972). Field observations of nearshore circulation and meandering currents. *J. Geophys. Res.* **77** (18): 3232–3247.

Stive, M. J. F. (1980). Velocity and pressure field of spilling breakers. *Proc. 17th Int. Conf. on Coastal Engineerng.* 547–566.

Stive, M. J. F. (1984). Energy dissipation in waves breaking on gentle slopes. *Coast. Eng.* **8** (2): 99–127.

Stive, M. J. F. and H. G. Wind (1982). A study of radiation stress and set-up in the nearshore region. *Coast. Eng.* **6** (1): 1–26.

Stive, M. J. F. and H.G. Wind (1986). Cross-shore mean flow in the surf-zone. *Coast. Eng.* **10** (4): 325–340.

Stokes, G. G. (1846). Report on recent researches in hydrodynamics. Report of the British Asscociation. Also in *Mathematical and Physical Papers (Collected)* (1880). 157–187.

Svendsen, I. A. (1984a). Wave heights and set-up in a surf-zone. *Coast. Eng.* **8** (4): 303–329.

Svendsen, I. A. (1984b). Mass flux and undertow in a surf-zone. *Coast. Eng.* **8** (4): 347–365.

Svendsen, I. A. (1987). Analysis of surf-zone turbulence. *J. Geophys. Res.* **92** (C5): 5115–24.

Svendsen, I. A. and R. S. Lorenz (1989). Velocities in combined undertow and longshore currents. *Coast. Eng.* **13** (1): 55–79.

Svendsen, I. A. and P. A. Madsen (1984). A turbulent bore on a beach. *J. Fluid Mech.* **148**: 73–96.

Svendsen, I. A., P. A. Madsen, and J. B. Hansen (1978). Wave characteristics in the surf-zone. *Proc. 16th Coastal Engineering Conf.* 520–539.

Svendsen, I. A. and U. Putrevu (1990). Nearshore circulation with 3-D profiles. *Proc.*

22nd Coastal Engineering Conf. 241–254.

Svendsen, I. A. and U. Putrevu (1993). Surf-zone wave parameters from experimental data. Coast. Eng. 19 (3&4): 283–310.

Svendsen, I. A. and U. Putrevu (1994a). Nearshore mixing and dispersion. Proc. Roy. Soc. Lond. A 445: 561–576.

Svendsen, I. A. and U. Putrevu (1994b). Velocity structure in IG-waves. Proc Int. Symp. Waves — Physical and Numerical Modelling, 21–24 August, Vancouver, Canada. 346–355.

Svendsen, I. A., H. A. Schäffer and J. B. Hansen (1987). The interaction between the undertow and boundary layer flow on a beach. J. Geophys. Res. 92 (C11): 11,845–11,856.

Symonds, G. and A. J. Bowen (1984). Interaction of nearshore bars with incoming wave groups. J. Geophys. Res. 89 (C2): 1953–1959.

Symonds, G., D. A. Huntley, and A. J. Bowen (1982). Two-dimensional surf-beat: Long wave generation by a time-varying break point. J. Geophys. Res. 87 (C1): 492–498.

Synolakis, C. E. and J. E. Skjelbreia (1993). Evolution of the maximum amplitude of solitary waves on plane beaches. J. Wtrwy., Port, Coast. and Oc. Eng. 119 (3): 323–342.

Synolakis, C. E. (1987). The runup of solitary waves. J. Fluid Mech. 185: 523–545.

Tallent, J. R., T. Yamashita, and Y. Tsuchiya (1989). Field and laboratory measurements of large scale eddy formation by breaking waves. Water Wave Kinematics, eds. A. Torum and O. T. Gudmestad, NATO ASI Series E: Applied Sciences, Vol. 178. 509–523.

Tang, E. and R. A. Dalrymple (1988). Rip currents and wave groups. Nearshore Sediment Transport, ed. R. J. Seymour, Plenum Publishing Corporation. 205–230.

Taylor, G. I. (1954). The dispersion of matter in a turbulent flow through a pipe. Proc. Roy. Soc. Lond. Series A 219: 446–468.

Thornton, E. B. (1970). Variation of longshore current across the surf-zone. Proc. 12th Coastal Engineering Conf. 291–308.

Thornton, E. B. (1979). Energetics of breaking waves within the surf-zone. J. Geophys. Res. 84 (C8) 4931–4938.

Thornton, E. B. and R. T. Guza (1982). Energy saturation and phase speeds measured on a natural beach. J. Geophys. Res. 87 (C12): 9499–9508.

Thornton, E. B. and R. T. Guza (1983). Transformation of wave height distribution. J. Geophys. Res. 88 (C10): 5925–5938.

Thornton, E. B. and R. T. Guza (1986). Surf-zone longshore currents and random waves: Field data and models. J. Phys. Oceanography 16 (7): 1165–1178.

Ting, F. C. K. and J. T. Kirby (1994). Observation of undertow and turbulence in a laboratory surf-zone. Coast. Eng. 24 (1): 51–80.

Trowbridge, J. H. and O. S. Madsen (1984). Turbulent wave boundary layers, Parts 1 and 2. *J. Geophys. Res.* **89** (C5): 7989–8007.

Tucker, M. J. (1950). Surf beats: Sea waves of 1 to 5 minute period. *Proc. Roy. Soc. Lond.* **A202**: 565–573.

Ursell, F. (1952). Edge waves on a sloping beach. *Proc. Roy. Soc. Lond.* **A214**: 79–97.

van Dongeren, A. R., F. E. Sancho, I. A. Svendsen, and U. Putrevu (1994). Quasi 3D modeling of infragravity waves. *Proc. 24th Int. Conf. Coastal Engineering.* 2741–2754.

van Dorn, W. G. (1976). Set-up and run-up in shoaling breakers. *Proc. 15th Int. Conf. Coastal Engineering.* 738–751.

Visser, P. J. (1982). The proper longshore current in a wave basin. Communications on Hydraulics, Department of Civil Engineering, Delft University of Technology, The Netherlands, Report 82-1. 86.

Visser, P. J. (1984). A mathematical model of uniform longshore currents and comparison with laboratory data. Communications on Hydraulics. Department of Civil Engineering, Delft University of Technology, The Netherlands, Report 84-2. 151.

Watanabe, A. (1982). Numerical models of nearshore currents and beach deformation. *Coast. Eng. Jpn.* **25**: 147–161.

Watson, G. and D. H. Peregrine (1992). Low frequency waves in the surf-zone. *Proc. 23rd Int. Conf. Coastal Engineering.* 818–831.

Watson, G., D. H. Peregrine, and E. F. Toro (1992). Numerical solution of the shallow water equations on a beach using the weighted average flux method. *Computational Fluid Dynamics* **1**: 495–502.

Whitford, D. J. (1988). Wind and wave forcing of longshore currents across a barred beach. Ph.D. Dissertation. Naval Postgraduate School, Monterey, USA. 205.

Wind, H. G. and C. B. Vreugdenhil (1986). Rip current generation near structures. *J. Fluid Mech.* **171**: 459–476.

Winer. H. S. (1988). Numerical modeling of wave-induced currents using a parabolic wave equation. Department of Coastal and Oceanographic Engineering, University of Florida, USA, UFL/COEL-TR/080.

Wright, L. D., J. Chappell, B. G. Thom, M. P. Bradshaw, and P. Cowell (1979). Morphodynamics of reflective abd dissipative beach and inshore systems: Southeastern Australia, *Marine Geology* **32**: 105.

Wright, L. D., R. T. Guza, and A. D. Short (1982). Dynamics of a high energy dissipative surf-zone. *Marine Geology* **45** (1&2): 41–62.

Wu, C.-S. and P. L-F. Liu (1985). Finite element modeling of nonlinear coastal currents. *J. Wtrwy., Port, Coast. and Oc. Eng.*, ASCE **111** (2): 417–432.

Wu, C.-S., E. B. Thornton, and R. T. Guza (1985). Waves and longshore currents: Comparison of a numerical model with field data. *J. Geophys. Res.* **90** (C3): 4951–4958.

Zelt, J. A. (1991). The run-up of non breaking and breaking solitary waves. *Coast. Eng.* **15** (3): 205–246.

PHYSICAL MODELING OF COASTAL PROCESSES

J. W. KAMPHUIS

1. Introduction

This paper describes physical modeling of coastal problems. Against a background historical perspective, the paper presents both the strengths and weaknesses of physical modeling and identifies the role of physical modeling in the 1990s alongside readily available hands-on computing power.

The paper is organized purposely at two levels. The expert in physical modeling can read the whole paper and find specific and detailed discussions on model scaling, using both equations and dimensional analysis in Secs. 4 to 7. The reader who is less informed on actual modeling details but is marginally involved (as a client or manager) or wants to gain a general understanding about modeling can skip these more technical sections and find a complete, more general discussion of modeling embodied in the remainder of the paper.

2. Background

Physical modeling of hydraulic phenomena is probably the *natural extension* of games children play when they are in or near water. Surface runoff during a rainstorm suddenly becomes a candidate for dam construction and channeling; canals are dug into a beach to supply water to a sand castle which simply must have a moat and the bathtub is the experimental basin of choice, where wave action, seiching, floatation (and sinking) of all kinds of vessels as well as rotational flow near the drain are extensively investigated.

Early work on physical modeling was no doubt a mere application of the simple rules learned as children. It was assumed that the same principles were valid in both the real world and the miniature world and geometric scaling (simply making the dimensions of the model consistent with the prototype) would have been the first introduction of rigor into modeling. If the model covered a large aerial extent, it would also have been necessary to cheat a little and increase the depth in the model in order for flow to take place at all (distortion).

It must soon have become obvious that simply obeying geometric scaling laws resulted in considerable confusion with respect to flow velocities (kinematic phenomena) and forces (dynamic phenomena).

Sextus Julius Frontinus (40–103 AD), who commissioned a hydraulic model of the Roman aqueduct system to determine where sand deposited in the system (Le Méhauté, 1990), must have faced both the necessity to distort the model and the puzzling aspects of velocity and time scales. Surely, early modelers simply did the best they could, using geometric scaling and then measured everything else. Sextus Julius would have found that sand would stop moving where fluid velocities decreased and must have recommended modifications to the system accordingly. Therefore, although the scaling of kinematic and dynamic phenomena was not clear to him and geometry was most likely distorted, he achieved his aim with this model.

We now have a much better understanding of the principles involved in fluid flow and the laws (rules) that bring about kinematic and dynamic similarity between model and prototype. The particular problems that perplex junior in the bathtub or may have confounded Sextus Julius perhaps are no longer of major concern to us. Does that mean that the answers provided now by physical models designed and operated in sophisticated laboratory facilities by modeling experts give satisfactory answers? Not always! As the models become more sophisticated, we tend to ask more complex questions and our answers continue to fall far short of our expectations. Thus, physical modeling techniques have indeed progressed in time, but the answers provided to the questions fall at least as far short as years ago. This means that common sense, intuition, and experience continue to be the most important ingredients of a successful model study and essentially, many physical models of complex problems are built simply as best we can, calibrated for some known typical conditions and then run for additional conditions of interest. Le Méhauté (1990) calls this "similitude by calibration". Thus, physical modeling continues to be an art as well as an ever more exact science (Hughes, 1993; Kamphuis, 1975 and 1982; Le Méhauté, 1990).[a]

---

[a]Common sense, intuition, experience, calibration, and the art of modeling are of course also relevant for numerical modeling.

As an example, consider the study of the sinking of the Ocean Ranger, a semi-submersible drilling rig which sank off Newfoundland in 1982 with the loss of 84 lives (Mogridge, 1985; Mogridge et al., 1986). The Royal Commission set up to investigate this disaster commissioned the National Research Council of Canada to reconstruct what happened and the main tool used was a hydraulic model. The "bathtub" was a multidirectional sea-keeping basin with a most sophisticated wave generation system and capability to subject the structure to wind forces. The most sophisticated measuring techniques were used. The water was relatively deep, the wave climate and the wind climate during the storm had both been measured a short distance from the site. This should have been an ideal study, a modeler's dream.

Testing showed, however that under the measured wind and waves, the structure did not experience stability problems; even large episodic waves, combining the worst of all cases, could not sink the Ocean Ranger model. Yet the prototype structure sank. During the testing, the modelers' careful observations did indicate some possible causes for the failure. It was concluded that the structure could capsize if its ballast water was shifted in a major way, changing the trim of the structure. But the balance of the prototype structure was carefully controlled by a complex set of pumps and valves, which had kept this structure upright for many years and in different environments. Thus, the modelers began to look for pump or valve failure prior to the sinking. The additional work carried out by the modelers and the Royal Commission concluded that a series of events began when large waves broke a small porthole in the ballast control room. This event did not lead directly to sinking the structure by the inflowing water; it created an electrical short circuit in a control panel, which in turn resulted in the control valves opening while indicator lights showed normal operating conditions. The combination of the open control valves and the faulty indicator lights resulted in the structure trimming forward. In time the structure listed so much that the waves washed over the deck, filling the chain lockers first and then the living quarters. The pumps, located in the stern, had been turned on manually by now, but they merely cavitated. It was under those conditions that the Ocean Ranger was finally capsized by the waves and this condition could also be simulated in the model.

The last part of this investigation is no longer science, it is pure ingenuity (or art or detective work). Does this remind you of junior trying to overturn the soap dish with a monster wave in the bathtub? Indeed it should. However, all the sophistication in scaling and modeling and the state-of-the-art equipment were necessary. Without it, the modelers would not have concluded that wind and wave action alone could not be responsible for the sinking. It permitted the modelers to discover, isolate, and concentrate on the real cause of the sinking and in fact piece together a very complex (and highly unlikely) chain of events.

Le Méhauté (1990) summarizes this when he says: "The use of scale modeling provides the **intuitive knowledge** gained by personally experiencing the elements. It is what permits the discovery of solutions that often seem like serendipity." Although graphical output from numerical models is often used in the same way, the physical model results have the distinct advantage of using actual physical similarity. Many nonlinear interactions and aspects which are not clearly described by the known equations are therefore modeled at least partially correctly and complex boundary conditions can be included without undue difficulty.

Before discussing physical modeling of coastal systems in detail, it may be helpful to identify the more common types of models. One broad differentiation is between a *design model* and a *process model* (Kamphuis, 1991). The design model simulates actual complex prototype situations in order to provide specific information that can be used in design or in retrospective study of failures. Breakwater stability studies, accretion, and erosion near harbor entrances and the Ocean Ranger model mentioned above are examples. The process model studies a particular physical process, such as how vortices near bedform ripples move sediment up into the water column or how wind waves influence diffusion. Boundary conditions and scales in these more abstract models are specified to minimize laboratory and scale effects.

Design models may be further subdivided into *long-term* and *short-term* design models. The short-term models simulate a prototype situation that takes place in hours or days, for example, one particular storm or series of storms. Both a breakwater stability study and the Ocean Ranger study would be short-term studies. Long-term studies refer to the classical hydraulic models which simulate response to fluid flow and wave climates which span months or years. Examples are models of shore morphology near structures, river meandering, and the development of inlets. These are the models that achieve similitude mainly by calibration (Le Méhauté, 1990).

In the following sections, the basics of empirical methods of determining model scales are first discussed. These methods are based on the assumption that equations as well as dimensionless functional relationships for fluid flow apply equally to a prototype and its model. It will be shown that it is not possible to satisfy all the required scaling relationships because of the model materials used and the laboratory space available. As a result most models contain *scale effects*, defined as nonsimilarity between the model and the prototype resulting from violation of basic scaling relationships.

Because the models are expected to simulate rather complicated prototype conditions with relatively simple modeling tools, models will also contain *laboratory effects*, nonsimilarity resulting from simplification of the prototype forcing and boundary conditions. For example, many years of prototype waves are modeled by repeating two or three typical wave conditions, and short-crested waves are modeled using a long-crested wave generator.

The implications of scale and laboratory effects on physical modeling are also presented. The rise and fall of the long-term design model is discussed and the future of physical modeling is shown to be an integration of process modeling and relatively simple computations. It is also shown that the trend away from the large design models requires a thorough understanding of modeling theory. It also needs improved modeling equipment and methodology, since process models are designed specifically to minimize scale and laboratory effects.

3. Model Scaling

The scaling of a physical model has been discussed in many references and only the most basic concepts will be repeated here. Langhaar (1951) points to the two basic methods of deriving scales when he says: "Dimensional analysis has been developed principally by British and American scientists. In continental Europe, model laws have been derived exclusively from the differential equations that govern the phenomena." He continues by pointing out that both methods have their strengths and limitations. Essentially that means both should be used to take advantage of the strengths of both.

A physical model simulates the prototype better than either equations or dimensional analysis and that is why physical models are used. Dimensional analysis describes a process in general form — physical models give detail. Physical models also automatically include many nonlinear and complex processes and boundary conditions that are not clearly understood and therefore cannot yet be expressed by the equations of a numerical model.

Strangely, both dimensional analysis and the use of equations have their adherents and considerable effort has been spent in the literature to show that one method is superior to the other. Le Mehauté (1990) states that dimensional analysis is a poor substitute for theory. Indeed it is, if the process is thoroughly understood and the theory describing it is well developed. But for most problems that are solved by physical models, the governing equations are only partially known and often based on simplifying assumptions; if all the equations and their interactions were known, a model would not be necessary. This means that for a better understanding, we need to rely, at least partly, also on dimensional analysis. Similarly, to argue that dimensional analysis should be used in order to understand the problem to the exclusion of what we already know from equations is incorrect. Langhaar (1951) points out that dimensional analysis is no substitute for equations. He states that its generality is both its strength and its weakness; with little effort, it provides a partial solution to almost any problem, but it never supplies a complete solution.

Most such partisan arguments arise because the use of a particular method is normally demonstrated by focusing on simple, familiar, fluid mechanics problems. This is necessary to keep examples short and understandable. But demonstrating how wonderful dimensional analysis is using flow over a weir or drag force on a sphere is not going to impress those who favor equations. If the process is well understood and the equations are well known, why go through the whole dimensional exercise? Therefore, any example developed in this review paper should not be so simple.

Many examples of coastal models could be developed here such as models of wave agitation in harbors, wave interaction with structures, or spreading of contaminants by currents and waves, and indeed, extensive examples may be found in the literature, such as Sharp (1981). Since the coastal mobile bed sediment transport and morphology model is perhaps the most difficult of all physical hydraulic models, it will be used here as an example. The following four sections are primarily presented to show the kind of reasoning required to achieve quantitatively successful models. They demonstrate the use of both equations and dimensional analysis methods and that we need both to succeed. They also show that all the methods fall far short of being simple, all-inclusive recipes for model design.

4. Scaling Using the Equations

To be able to use equations, they must be correctly formulated and dimensionally consistent. Then, to derive scaling relationships really requires collecting all the equations that apply to the problem. Each equation will highlight certain aspects and can be used to derive particular scales. The assumption is that each equation applies to the model as well as to the prototype. There are two basic methods in use: the model scales may be derived directly from the equations as is done below, or some authors prefer to first make the equations dimensionless, for example, Langhaar (1951), Shen (1990a), and Hughes (1993).

Let us consider the derivation of scales for the fluid portion of a coastal morphology model. The basic fluid flow in a coastal problem may be described by the continuity equation and the well-known Navier Stokes equations, given in differential form as

$$\frac{\partial u}{\partial x} + \frac{\partial v}{\partial y} + \frac{\partial w}{\partial z} = 0 \tag{1}$$

$$\frac{\partial u}{\partial t} + u\frac{\partial u}{\partial x} + v\frac{\partial u}{\partial y} + w\frac{\partial u}{\partial z} = -\frac{1}{\rho}\frac{\partial p}{\partial x} + \nu\left(\frac{\partial^2 u}{\partial x^2} + \frac{\partial^2 u}{\partial y^2} + \frac{\partial^2 u}{\partial z^2}\right)$$
$$- \left(\frac{\partial}{\partial x}(\overline{u'^2}) + \frac{\partial}{\partial y}(\overline{u'v'}) + \frac{\partial}{\partial z}(\overline{u'w'})\right) \tag{2}$$

$$\frac{\partial v}{\partial t} + u\frac{\partial v}{\partial x} + v\frac{\partial v}{\partial y} + w\frac{\partial v}{\partial z} = -\frac{1}{\rho}\frac{\partial p}{\partial y} + \nu\left(\frac{\partial^2 v}{\partial x^2} + \frac{\partial^2 v}{\partial y^2} + \frac{\partial^2 v}{\partial z^2}\right)$$
$$- \left(\frac{\partial}{\partial x}(\overline{u'v'}) + \frac{\partial}{\partial y}(\overline{v'^2}) + \frac{\partial}{\partial z}(\overline{v'w'})\right) \tag{3}$$

$$\frac{\partial w}{\partial t} + u\frac{\partial w}{\partial x} + v\frac{\partial w}{\partial y} + w\frac{\partial w}{\partial z} = -\frac{1}{\rho}\frac{\partial p}{\partial z} + g + \nu\left(\frac{\partial^2 w}{\partial x^2} + \frac{\partial^2 w}{\partial y^2} + \frac{\partial^2 w}{\partial z^2}\right)$$
$$- \left(\frac{\partial}{\partial x}(\overline{u'w'}) + \frac{\partial}{\partial y}(\overline{v'w'}) + \frac{\partial}{\partial z}(\overline{w'^2})\right). \tag{4}$$

Here, u, v, and w are the velocity components in the x, y, and z directions, t is time, p is pressure, g is gravitational acceleration, ρ is the density of the water, ν the kinematic viscosity of the water, and u', v', and w' are the turbulent fluctuations of the velocities.

Let us define the scale of a quantity X as

$$n_X = \frac{X_P}{X_M} \tag{5}$$

where the subscripts P and M refer to prototype and model respectively. For Eqs. (1) to (4) to apply equally to both prototype and model, each term must be the same in prototype and model. Dividing each term of Eq. (4) as written for the prototype by each term as written for the model gives dimensionless ratios that result in the following scaling laws:

$$\frac{n_w}{n_t} = \frac{n_u n_w}{n_x} = \frac{n_v n_w}{n_y} = \frac{n_w^2}{n_z} = \frac{n_p}{n_\rho n_z} = n_g$$

$$= \frac{n_\nu n_w}{n_x^2} = \frac{n_\nu n_w}{n_y^2} = \frac{n_\nu n_w}{n_z^2} = \frac{n_{u'} n_{w'}}{n_x} = \frac{n_{v'} n_{w'}}{n_y} = \frac{n_{w'}^2}{n_z}. \qquad (6)$$

From this and other similar relationships derived from Eqs. (1) to (4), it is clear that all distance scales must be equal and all the velocity scales must be equal. Dividing the scale for an inertia term, such as term 4 of Eq. (6), by the gravity term (term 6) yields

$$\frac{n_w^2}{n_g n_z} = 1. \qquad (7)$$

Equation 7 may be recognized as a Froude number scale; it states that the Froude number must be the same in the model and the prototype. Similarly dividing the viscosity term (term 9) into term 4 yields the recognizable Reynolds number scale:

$$\frac{n_w n_z}{n_\nu} = 1. \qquad (8)$$

If the gravitational acceleration is the same for the model and prototype, and the model fluid has roughly the same viscosity as the prototype fluid (usually both are water), then Eq. (7) says that $n_w = \sqrt{n_z}$ while Eq. (8) dictates that $n_w \approx 1/n_z$. Since this can only be true for $n_\nu = n_z = 1$, this means that the prototype is the only valid model. So if we want to design a scale model, we must use a fluid with a correctly scaled viscosity (very difficult) or change g for the model (such as in a centrifuge model) or we must decide, using physical reasoning, whether gravity or viscosity is the least important parameter for the problem at hand and ignore its effect.

For the usual hydraulic models, the scaling of viscosity and gravity is not practical and gravity effects are considered to be more important. This means that the Froude number will be represented correctly (Eq. (7)) while the incorrect representation of the viscosity terms will cause the model not to behave exactly like the prototype. This difference between the model and the prototype, resulting from not being able to adhere to certain scaling criteria is usually referred to as *scale effect* and it is imperative that the modeler understands the consequences each time an improper scaling is introduced. For example, not satisfying viscosity requirements may not be serious as long as the flow in the model is always rough-turbulent, but for coastal models in which the orbital velocity goes through zero twice in every wave cycle, this could have some serious implications.

For hydraulic models in which the viscosity effect is neglected and $n_g = 1$ (the gravitational acceleration in the model is the same as in prototype), the following basic scales may now be derived from Eqs. (1) to (4):

$$n_x = n_y = n_z = n_{(p/\rho g)} = n$$
$$n_u = n_v = n_w = n_{u'} = n_{v'} = n_{w'} = n_t = \sqrt{n}. \tag{9}$$

Here, n denotes the geometric scale. To learn more about the waves in a coastal model, let us first examine the depth integrated version of Eqs. (1) to (4) (Phillips, 1977).

$$\frac{\partial U}{\partial t} + U\frac{\partial U}{\partial x} + V\frac{\partial U}{\partial y} = -g\frac{\partial \eta}{\partial x} - \frac{\tau_{bx}}{\rho d} - \frac{1}{\rho d}\left(\frac{\partial S_{xx}}{\partial x} + \frac{\partial S_{xy}}{\partial y}\right) + \frac{\tau_{ly}}{\rho d} \tag{10}$$

$$\frac{\partial V}{\partial t} + U\frac{\partial V}{\partial x} + V\frac{\partial V}{\partial y} = -g\frac{\partial \eta}{\partial y} - \frac{\tau_{by}}{\rho d} - \frac{1}{\rho d}\left(\frac{\partial S_{xy}}{\partial x} + \frac{\partial S_{yy}}{\partial y}\right) + \frac{\tau_{lx}}{\rho d} \tag{11}$$

$$\frac{\partial \eta}{\partial t} + \frac{\partial(Ud)}{\partial x} + \frac{\partial(Vd)}{\partial y} = 0 \tag{12}$$

where U and V are the depth averaged velocities in the x and y directions, normally defined as shore perpendicular and shore parallel respectively, η is the wave set-up, τ_b is the shear stress on the bottom of the control volume, τ_l are the lateral shear stresses on the sides of the control volume, S is the radiation stress of the waves, and d is the depth of the water.

Additional scales may now be derived from Eqs. (10) to (12).

$$n_\eta = n; \quad n_\tau = n_\rho n; \quad n_U = n_V = \sqrt{n}; \quad n_S = n_\rho n^2. \tag{13}$$

Further insight into the wave kinematics may be derived, for example, from the small amplitude expression for wave dispersion

$$L = \frac{gT^2}{2\pi} \tanh\left(\frac{2\pi d}{L}\right) \tag{14}$$

and from the expression for horizontal orbital velocity component

$$u_w = \frac{\pi H}{T} \frac{\cosh\frac{2\pi}{L}(d+z)}{\sinh\frac{2\pi d}{L}} \cos 2\pi\left(\frac{x}{L} - \frac{t}{T}\right) \tag{15}$$

where L is the wave length, T the wave period, and H is the wave height. Writing these equations for prototype and dividing by the same equation for the model yields

$$n_L = n_T^2 n_{\tanh}; \quad n_{u_w} = \frac{n_H}{n_T} \frac{n_{\cosh}}{n_{\sinh}} n_{\cos}. \tag{16}$$

Obviously the mathematical expressions tanh, cosh, sinh, and cos must produce the same results in the model and prototype. This means their scales must equal 1, which in turn means that their arguments must have the same values in the model and prototype. This can only be achieved if

$$n_L = n_d = n_z = n_x = n_T^2; \quad n_t = n_T. \tag{17}$$

It would appear reasonable to assume that n_H must be equal to n_η and n_d and therefore equal to the geometric scale. Note that this could have been derived by introducing other equations, but here we actually use a crude form of dimensional analysis. Finally, because of scale effects (discussed later) and errors in τ and S, we may assume that $n_\rho = 1$, even when seawater is modeled by fresh water. This results in one possible scaling combination based on modeling the Froude criterion correctly:

$$n_x = n_y = n_z = n_{(p/\rho g)} = n_d = n_\eta = n_H = n_L = n_\tau = n$$
$$n_u = n_v = n_w = n_U = n_V = n_{u'} = n_{v'} = n_{w'} = n_{u_w} = n_t = n_T = \sqrt{n} \tag{18}$$
$$n_S = n^2.$$

Other equations could be introduced to represent other wave related parameters. As long as one introduces the correct equations to describe each aspect of the model, it will be possible to derive the pertinent scaling relationships. Examples of other equations and their conversion into scaling relationships may be found in Hughes (1993).

The selection of the relevant equations is where the real ingenuity and thinking of the modeler become important. Which equations should be used, which are the most important, which equations are redundant? Note that if an equation is not included in the analysis, some important relationships may be completely missed.

Often, particularly for long term design models, it is not possible to use the same horizontal and vertical scales and the model must be *geometrically distorted*. The ratio of the aerial extent of the prototype to be modeled over the size of the laboratory determines the horizontal scales. This is usually of the order of 100 or greater (10 meters of model floor space represents 1 km of prototype or more). If this same scale were used for the vertical scale, some model depths, wave heights, bedforms, etc. would become too small. In such cases, it is usual to choose the depth scale to be smaller than the horizontal scales (a depth scale of 30 to 50 is usually adopted). Denoting the vertical scale now by n, we define geometric distortion as

$$N = \frac{n_x}{n} = \frac{n_y}{n}. \qquad (19)$$

It is usual in coastal models to satisfy the Froude criterion for all velocities, that is, all velocity scales are kept the same and equal to $\sqrt{n}$. Introducing the general definitions for current velocities in the (distorted) horizontal directions

$$U = \frac{dx}{dt}; \qquad V = \frac{dy}{dt} \qquad (20)$$

yields

$$n_U = \frac{n_x}{n_t} = \frac{nN}{n_t} = n_V = \sqrt{n} \quad \rightarrow \quad n_t = N\sqrt{n} \qquad (21)$$

which means that the time scale for currents is now distorted. Similarly, Eqs. (14) and (15) show that L remains related to the depth scale. This simply means that there are fewer waves per unit horizontal distance in a distorted meal.

Denoting the time scales for the currents by subscript c we can now write for a geometrically distorted model based on Froude scaling:

$$\begin{aligned}
&n_z = n_{(p/\rho g)} = n_d = n_\eta = n_H = n_L = n_\tau = n \\
&n_x = n_y = Nn \\
&n_u = n_v = n_U = n_V = n_w = n_{u'} = n_{v'} = n_{w'} = n_{u_w} = \sqrt{n} \\
&n_t = n_T = \sqrt{n} \\
&n_{t_c} = n_{T_c} = N\sqrt{n} \\
&n_S = n^2.
\end{aligned} \qquad (22)$$

As with ignoring the Reynolds number earlier, geometric distortion of the model introduces further scale effects. For example, from Eqs. (1) to (4), it is clear that $\partial u/\partial z$, $\partial v/\partial z$, $\partial p/\partial x$, $\partial p/\partial y$, the viscosity term and the turbulence term are all distorted, in other words, geometric distortion of the model could have serious consequences and the modeler must be aware of these. Also, because the number of waves per unit horizontal distance is now incorrect, wave diffraction equations will show that diffraction is no longer correct and hence a wave model involving wave diffraction must be undistorted.

Note again that Eqs. (22) were arrived at by some dimensional reasoning (such as, that the Froude number must be constant). It is now time to look at dimensional analysis in more detail.

5. Dimensional Analysis

Dimensional analysis is primarily known as a tool that is used for scaling physical models. It is, however, much more general than that. Since it can be used to describe any physical process, it is also an excellent tool to provide understanding and organization for field experiments and numerical models.

Dimensional analysis is based on *theory of dimensions* which assumes that all fundamental equations are *dimensionally homogeneous* and that all physical quantities may be expressed as a power product of a number of independent basic entities used internationally as units of measurement. Hydraulics problems may usually be expressed in three base units: mass $[M]$, length $[L]$, and time $[T]$, where the $[\,]$ denotes "units of". All other quantities then have units derived from these; velocity has units $[M^0 L^1 T^{-1}]$, force $[M^1 L^1 T^{-2}]$, and density $[M^1 L^{-3}]$. This means that any physical quantity can be expressed in dimensionless form by being dividing by some power product of $[M]$, $[L]$, and $[T]$. More importantly, any physical quantity can also be made dimensionless by being divided by a power product of any other three independent physical quantities (to be independent, $[M]$, $[L]$, and $[T]$ must appear at least once among the three derived units). A dimensional expression with n independent variables (a_i)

$$A = f(a_1, a_2, a_3, a_4, \ldots, a_{n-1}, a_n) \qquad (23)$$

may now be reduced to a dimensionless expression with $(n-r)$ dimensionless independent variables (X_i)

$$\Pi_A = \phi_A(X_1, X_2, X_3, \ldots, X_{n-r}) \qquad (24)$$

where r is the number of base units pertinent to the problem (usually three: $[M]$, $[L]$, and $[T]$). There are some basic rules to dimensional analysis and details of the method have been discussed in many references, for example, Hughes (1993), Ivicsics (1975), Kamphuis (1975, 1985, and 1991), Langhaar

(1951), and Yalin (1971). Because dimensional analysis works with functional relationships such as Eqs. (23) and (24), it says nothing about the actual relationships, only that their dimensions must be correct.

For the fluid phase of a coastal model we could say that

$$A = f(H, T, d, k, \rho, \mu, g, x, y, z, t) \tag{25}$$

where A is any dependent variable, for example, wave orbital velocity, wave length, etc., k is the bottom roughness, and μ is the dynamic viscosity of the fluid. The list of independent variables must include all the variables pertinent to the dependent variable and should exclude all variables that are not independent or are irrelevant. This is a rather daunting assignment, particularly for the "uninitiated".

If we choose u_w to be the dependent variable in Eq. (25), we can state that the following power product must be dimensionless:

$$\Pi = u_w^{\beta_u} H^{\beta_H} T^{\beta_T} d^{\beta_d} k^{\beta_k} \rho^{\beta_\rho} \mu^{\beta_\mu} g^{\beta_g} x^{\beta_x} y^{\beta_y} z^{\beta_z} t^{\beta_t} . \tag{26}$$

Since each parameter (including the final dimensionless product) is a power product of $[M]$, $[L]$, and $[T]$, we can set up a matrix for the exponents of $[M]$, $[L]$, and $[T]$.

	u_w	H	T	d	k	ρ	μ	g	x	y	z	t
M	0	0	0	0	0	1	1	0	0	0	0	0
L	1	1	0	1	1	-3	-1	1	1	1	1	0
T	-1	0	1	0	0	0	-1	-2	0	0	0	1

For Π in Eq. (26) to be dimensionless the exponents of $[M]$, $[L]$, and $[T]$ must be zero. Therefore

$$\beta_\rho + \beta_\mu = 0$$
$$\beta_{u_w} + \beta_H + \beta_d + \beta_k - 3\beta_\rho - \beta_\mu + \beta_g + \beta_x + \beta_y + \beta_z = 0 \tag{27}$$
$$\beta_{u_w} - \beta_T + \beta_\mu + 2\beta_g - \beta_t = 0.$$

These three equations in 12 unknowns will yield an *infinite number of solutions*. Experience with the problem at hand will help to select those solutions that make most physical sense. Langhaar (1951) and Hughes (1993) demonstrate this approach.

Yalin (1971) and Kamphuis (1975, 1985, and 1991) follow a slightly more organized approach to determine a first reasonable set of dimensionless variables. First, they are careful to separate the dependent variable by keeping it on the left side of the equation as in Eq. (25). This is simply a reminder that helps to prevent *spurious correlation* (Benson, 1965; Kamphuis and Yalin, 1971). Then they choose three variables called *repeaters*, which they will use repeatedly to form the dimensionless ratios. The actual choice of the repeaters does not matter much as long as they are independent. Two useful rules of thumb for selecting the repeaters are given below. The resulting ratios are all dimensionless. Therefore, a power product of any combination of them will still be dimensionless. Thus, the initially derived dimensionless ratios can be recombined at any time to form new dimensionless ratios that make most physical sense.

The first guide in selecting repeaters is to choose three independent parameters which are most difficult to vary, so that the readily varied parameters are assigned a dimensionless ratio of their own. For Eq. (25), the repeaters would be ρ, μ, and g. These repeaters would, however, result in rather awkward dimensionless ratios. A second good rule is to simplify the process by making at least one repeater a representative length. Using ρ, d, and g as repeaters, Eq. (25) becomes

$$\Pi_A = \phi \left(\frac{H}{d}, T\sqrt{\frac{g}{d}}, \frac{k}{d}, \frac{\mu}{\rho d\sqrt{gd}}, \frac{x}{d}, \frac{y}{d}, \frac{z}{d}, t\sqrt{\frac{g}{d}} \right). \quad (28)$$

Note that in this dimensionless notation, the number of independent variables has decreased by three. Eleven dimensional variables in Eq. (25) have become 8 dimensionless variables in Eq. (28); there are no dimensionless ratios representing the repeaters. Are these the correct ratios? Here, the detailed knowledge by the modeler about the physics of the problem becomes important. For example, the first ratio $X_1 = H/d$ would suffice if a process near breaking is described, but in deep water, (H/L) would be better or even the related (H/gT^2). This latter parameter could be obtained by replacing the first ratio $X_1 = (H/d)$ by X_1/X_2^2. It is also usually better to relate the horizontal distances x and y to a typical parameter for horizontal distances. For waves this would obviously be a wave length. Since wave length is not independent from the other parameters in Eq. (25) (it is completely defined by H, T, and d), it was not included there.

If the dependent variable A in Eq. (25) were chosen to be L, then Eq. (28) would yield $\Pi_L = L/d$. Multiplying power products of this ratio into some of the right-side dimensionless ratios of Eq. (28) and replacing X_8 by $X_8 X_2^{-1}$ results in a much more useful function for deeper water:

$$\Pi_A = \phi_A \left(\frac{H}{L}, \frac{L}{gT^2}, \frac{k}{d}, \frac{\mu}{\rho d \sqrt{gd}}, \frac{x}{L}, \frac{y}{L}, \frac{z}{d}, \frac{t}{T} \right). \qquad (29)$$

Note that throughout this manipulation, the number of independent dimensionless variables has remained the same; the new dimensionless products always replace only one of the original products. Equation (29) is not the only valid set of dimensionless parameters. The parameters to be used depend very much on the problem at hand. For example, with x as the cross-shore direction and y as the alongshore direction, then X_5, X_6, and X_7 could be recombined to form $m = z/x$ (beach slope), $\alpha = x/y$ (incident wave angle), and z/d (relative depth). Replacing X_4 by $X_4^{-1} X_1^{1/2}$ to yield a more common Reynolds number and choosing u_w as the dependent variable, Eq. (29) becomes

$$\frac{u_w}{\sqrt{gd}} = \phi_{u_w} \left(\frac{H}{L}, \frac{L}{gT^2}, \frac{k}{d}, \frac{\sqrt{gH}d}{\nu}, m, \alpha, \frac{z}{d}, \frac{t}{T} \right). \qquad (30)$$

For the dimensionless relationships to be most useful, the modeler needs to have a thorough understanding of what exactly needs to be modeled, otherwise the resulting relationships will contain less-meaningful, dimensionless parameters. Note that the left side of Eq. (30) is a Froude number and that the viscosity term on the right side is a Reynolds number.

6. Scaling from Dimensions

Model scaling using dimensions is based on essentially the same principle as with the equations: we assume that the functions apply both to the prototype and the model. However, we do not know the actual detailed shape of the function ϕ, and therefore, to make $\phi_M = \phi_P$, all the independent dimensionless variables in the model must be the same as in the prototype. In other words, the scales for each of the dimensionless variables must equal one. From Eq. (30), it is first noticed, as with the scaling from equations discussed earlier, that in the usual situation where $n_g = 1$ and $n_\nu \approx 1$, it is not possible to satisfy both the Reynolds and Froude number criteria. Decisions must be made, based on the physical details of the problem, as to which is more important or how the whole problem can best be split up into (time or geometric) segments in which one of the two effects becomes predominant. An interesting discussion of this may be found in Schuring (1977). Usually for wave and current models the Reynolds number criterion is relaxed and the following scales may be derived from Eq. (30):

$$n_H = n_d = n_L = n_k = n_x = n_y = n_z = n$$
$$n_{u_w} = n_T = n_t = \sqrt{n} \qquad (31)$$
$$n_m = n_\alpha = 1.$$

It is also possible to introduce other dependent variables into Eq. (25), to derive dimensionless relationships for these. If we do this for p, u, v, U, V, w, u', v', w', S, η, and τ, we would end up with exactly the same scales as in Eq. (18), and if we introduce geometric distortion into the model, the end result would be the same as Eq. (22).

One complication that almost always occurs is that the bottom roughness, k, cannot be scaled down correctly by the model scale n. For a fixed bed model, the bed would need to be impossibly smooth, and for a mobile bed model, the bedform size does not scale down by a factor n and scaling the sediment itself down by n would cause the sediment to go into suspension for all prototype sediments finer than gravel. This means bottom roughness is almost always distorted. Roughness distortion may be defined as

$$N_k = \frac{n_k}{n_d} = \frac{n_k}{n} < 1. \qquad (32)$$

The implications of this will be discussed later.

7. Scaling from Both Dimensions and Equations

It was seen that dimensions and equations yielded identical scales for a physical model representing the fluid phase of a coastal model. Since it is simpler to think of all the pertinent variables that affect a problem than to think of all the equations that might possibly have an influence, it is probably easier and somewhat more methodical to begin with dimensional analysis and supplement it with known equations. This approach is followed below in determining scales required for the mobile bed portion of the coastal model.

Models involving sediment transport and bathymetry change normally involve both bed load and suspended load as well as areas of erosion and accretion. Each of these aspects has different requirements, and examples of how to determine scales and scale effects will now be discussed.

7.1. Shear stress modeling

A model that simulates erosion must model initiation of motion correctly. For many problems, it is possible to assume that transport is solely a function of bottom shear stresses. Such a model would simulate erosion offshore of the breaking zone. Let us concentrate on the sediment rather than on the whole problem, i.e., we assume that the shear stress can be separately related to the wave and current motion. In that case, the following dimensional relationship may be written

$$A = f(\rho, \mu, g, D, v_*, \rho_s, a_B) \tag{33}$$

where D is the grain size, v_* the shear velocity $\{= (\tau_0/\rho)^{1/2}\}$, τ_0 the bottom shear stress, ρ_s the sediment density, and a_B the wave orbital amplitude at the bottom. Using the same rules of thumb as before, we select ρ, g, and D as repeaters and obtain

$$\Pi_A = \phi_A \left(\frac{\mu}{\rho D \sqrt{gD}}, \frac{v_*}{gD}, \frac{\rho_s}{\rho}, \frac{a_B}{D} \right). \tag{34}$$

Although $\sqrt{(gD)}$ has the units of velocity, it does not actually represent a velocity like, for example, $\sqrt{(gd)}$ and we will replace it by v_*, the only velocity in Eq. (33). Further, assuming that initiation of motion always takes place under water (which makes $(\rho_s - \rho)$ important, rather than ρ_s), Eq. (34) may be reformulated as

$$\Pi_A = \phi_A \left(\frac{v_* D}{\nu}, \frac{\rho v_*^2}{(\rho_s - \rho)gD}, \frac{\rho_s}{\rho}, \frac{a_B}{D} \right) = \phi_A(R_*, M_*, \rho_*, G_*) \tag{35}$$

where R_* is the grain size Reynolds number, M_* the mobility number, ρ_* the relative density of the sediment, and G_* the geometric link between the waves and the sediment. In the following, $\{(\rho_s - \rho)/\rho\}$ will be identified by Δ. The first and second ratios form the axes of the Shields diagram. The second ratio could be expressed as excess mobility over a critical value for large grain sizes. All these ratios must be the same in model and prototype. This will be discussed later, but first let us examine shear stress.

Shear stress can best be related to wave and current action using theoretical or empirical equations. From these Kamphuis (1975, 1985, 1991) proposes three different shear stress scales which represent the *actual* shear stress under waves (τ_w) and under currents (τ_c), and the *required* shear stress to obtain the correct flow patterns (τ_f).

$$n_{\tau_w} = n N_k^{3/4}; \quad n_{\tau_c} = n N_k^{1/4}; \quad n_{\tau_f} = n N^{-1}. \tag{36}$$

For a flat bed (no bedform), k is a function of grain size as shown, for example, in Kamphuis (1974). When bedform is present, bottom shear is a function of skin friction along the grains and form drag resulting from the bedform geometry. Usually skin friction is negligible and k may be related directly to the bedform height.

At this point we will assume that we are concerned with waves only. Assuming $n_\rho = 1$ as before results in

$$n_{v_*} = n^{1/2} N_k^{3/8}. \tag{37}$$

The only model that would be able to satisfy Eq. (35) completely would be the prototype itself. The best any scale model can do is to simulate M_*, ρ_*, and G_* correctly. This would mean, using Eqs. (32), (35), and (37):

$$n_{\rho_s} = n_\rho; \quad n_D = n_{a_B} = n \quad \rightarrow \quad n_{v_*} = n^{1/2}; \quad n_{R_*} = n^{3/2} n_\nu^{-1} \simeq n^{3/2}. \tag{38}$$

Usually, such a model is not possible because the grain size needs to be scaled down by the model scale. Unless the prototype grain size is large (gravel), the model grain size becomes so fine that the material will go completely into suspension. Because R_* is not modeled correctly, a Reynolds number scale effect will be present (model Reynolds numbers will be much smaller than required). Detailed discussions of the various scale effects may be found in Kamphuis (1975, 1985, and 1991) and will not be repeated here.

When the grain size cannot be scaled down by the model scale, the G_* condition is violated and this again has serious scale effect implications which vary with location in the model. In addition, for the case of a flat bed,

$$N_k = \frac{n_k}{n} = \frac{n_D}{n} \quad \rightarrow \quad n_{v_*} = n^{1/2} N_k^{3/8} = n^{1/8} n_D^{3/8}. \tag{39}$$

For correct Reynolds number scaling,

$$n_{R_*} = 1 \quad \rightarrow \quad n_{v_*} n_D n_\nu^{-1} \simeq n^{1/8} n_D^{11/8} = 1 \quad \rightarrow \quad n_D = n^{-1/11} \tag{40}$$

which would mean that the model grain size would be somewhat larger than the prototype. This is not a practical solution because the model values of M_* and G_* would be much too small.

For correct mobility number scaling and using Eq. (39),

$$n_{v_*}^2 = n_\Delta n_D = n^{1/4} n_D^{3/4} \quad \rightarrow \quad n_D = n \, n_\Delta^{-4}. \tag{41}$$

If sand is used in the model, $n_\Delta \approx 1$. In that case n_D would be approximately equal to n, which is normally not possible as shown before. Thus, mobility number scaling can only be achieved with *lightweight material*. Hughes (1993) provides a table of possible lightweight materials. For bakelite or coal, $\rho_* = 1.4$, $n_\Delta = 4$, and for a model with $n = 50$, $n_D = 0.19$. If $D_p = 0.25$ mm, D_m would need to be 1.31 mm, which is reasonable. For polystyrene, however, $\rho_* = 1.05$, $n_\Delta = 33$ and for a model with $n = 50$, $n_D = 0.00042$ which is not practical.

One usual modeling ploy is to satisfy both the Reynolds and mobility criteria, thereby satisfying the Shields criterion for initiation of motion. This combines Eqs. (40) and (41):

$$n_{v_*} n_D \simeq 1; \quad n_{v_*}^2 = n_\Delta n_D = 1 \quad \rightarrow \quad n_D = n_\Delta^{-1/3}; \quad n_{v_*} = n_\Delta^{1/3} \qquad (42)$$

and, for example, for $n = 50$ and a prototype grain size of 0.25 mm, this would mean a model grain size of 0.40 mm for coal and bakelite and 0.80 mm for polystyrene.

The shear velocity defined by Eq. (42) is the required shear velocity to make the Shield criterion modeling work. The actual shear velocity scale is, however, as specified by Eq. (39) and it will be equal to $n_\Delta^{1/3}$ only if $n = n_\Delta^{11/3}$. This would result in model scales of about 160 for coal and bakelite and 370,000 for polystyrene. Both these would not be practical.

When we use smaller values of n, the shear velocity scale defined by Eq. (39) is smaller than $n_\Delta^{1/3}$, which means that model shear stress is too large. Such incorrect shear stress may not be serious if the purpose of the model is to evaluate bulk sediment transport processes. For detailed study, however, such as scour and deposition near structures, the scour holes would be too deep and the shoals would not build up high enough. Interpretation of such model results is very difficult.

The large grain sizes in lightweight models will furthermore result in relatively too great a porosity in the model. Finally, the ρ_* criterion is violated. This will result in a number of other serious scale effects (Kamphuis, 1975, 1985, and 1991). The conclusion has to be that lightweight models are not very practical for coastal research.

If the serious scale effects of using lightweight material are to be avoided, there is an alternative: use sand as the modeling material. Although some recipes exist to define model particle size such as Noda (1972) and Kriebel et al. (1986) who both find $n_D \approx \sqrt{n}$, it is normally best to use the smallest sand grain size that will not go into suspension in the model. This is of the order of 0.1 mm and it determines n_D. For example, for $n = 50$, $n_\Delta = 0.96$, and $D_p = 0.25$ mm, $n_D = 2.5$ which means that for a flat bed,

$$n_{v_*} = n^{1/8} n_D^{3/8} = 2.3 \quad \to \quad n_{R_*} \simeq n^{1/8} n_D^{11/8} = 5.7$$
$$n_{M_*} = n^{1/4} n_D^{-1/4} n_\Delta^{-1} = 2.2; \quad n_{\rho_*} = 1; \quad n_{G_*} = n n_D^{-1} = 20. \tag{43}$$

The model mobility number is only slightly smaller than required, and because most sediment transport processes are highly dependent on mobility number, this is good. There will, however, be scale effects, particularly because of the improper scaling of G_*, which causes the grain size to be much too large in relation to the driving forces of the waves.

The above discussion is all related to models (and prototypes) with a flat bed. Of course, many prototypes and almost all sand models exhibit *bedform*. Bedform length is related very closely to wave orbital size (a_B). Since equilibrium bedform has an almost constant steepness regardless of size, the bedform height δ, for equilibrium bedform is also related to a_B (Mogridge and Kamphuis, 1972; Kamphuis, 1988). Since bottom friction consists of skin friction and form drag and since the latter is usually more important, k becomes a simple function of bedform height δ (Kamphuis, 1988) and

$$n_\delta = n \quad \to \quad n_{v_*} = n^{1/2} N_\delta^{3/8} = n^{1/2}. \tag{44}$$

This would result in simple scaling, since now there is no roughness distortion. However, from experience we know that equilibrium bedform in the prototype is rare and that in models the bedform sometimes washes out, particularly in the breaking zone. This means n_δ will vary over the model and the modeler has to be very careful to recognize and understand the additional scale effects from roughness distortion.

Until now we have only considered waves. When currents are important also, the other shear stress scales come into play. This makes the model even more complicated and, based on available information, a decision is usually made whether waves or currents dominate the process studied.

7.2. Fall velocity modeling

The above discussion was only concerned with underwater sediment transport caused by wave-generated shear stress. If the sediment phase of the model must simulate suspended load and deposition of sediment, the suspending mechanism and the fall velocity ω become important.

There are several theories to describe how material becomes suspended and Fredsøe and Deigaard (1992) present a good summary. Scales can be derived from the equations of turbulent eddy viscosity, mixing length, and the parameters in the various turbulence models. The net result is

$$n_{v_*} = n^{1/2}. \tag{45}$$

Since fall velocity may be expressed as

$$\omega = f(\rho, \mu, D, \rho_s, g) \tag{46}$$

it could not simply be included in the independent variable list of Eq. (33).[b] The practical importance of fall velocity is that it determines the distance that a suspended sediment particle moves in a certain time; it determines where a particle settles. For a distorted model, this means horizontal distance should be used to define the settling patterns of the suspended sediment. This requires that we return to Eq. (25) in which horizontal distance x and y occur. Replacing k (which is a function of D) by fall velocity ω in Eq. (25) results in a new dimensionless scaling relationship:

$$\omega_* = \frac{x}{\omega T} \quad \rightarrow \quad n_\omega = n^{1/2} N. \tag{47}$$

7.3. Breaking zone

The discussion in the above two subsections was concerned with sediment motion in the relatively gentle environment offshore of the breaking zone. In the breaking zone, bottom shear is not the prime moving force, but the wave energy dissipation rate and wave momentum (radiation stress) cause the sediment transport. For a breaking zone model to simulate erosion under water, the following example function may be derived from Eq. (25) after replacing k with D and ρ_s:

$$\Pi_A = \phi_A \left(\frac{H_b}{d_b}, \frac{H_b}{L_b}, m, \alpha_b, \frac{t}{T}, \frac{\sqrt{gH_b}D}{\nu}, \frac{H_b}{\Delta D}, \frac{\rho_s}{\rho}, \frac{H_b}{D} \right). \tag{48}$$

[b]ω was incorrectly included as an independent variable in Kamphuis (1991).

The last four ratios in this equation are other forms of R_*, M_*, ρ_*, and G_*. Many of the arguments presented in the above subsections are also valid for the breaking zone. Lightweight materials, for example, are equally unsuitable as described by Kamphuis (1975, 1985, and 1991). All the scales of Eqs. (22) and (31) apply and for the example discussed earlier, using the smallest sand that does not go into suspension, for $n = 50$, $n_\Delta = 0.96$, $D_p = 0.25$ mm, $n_D = 2.5$:

$$n_{R_*} \simeq n^{1/2} n_D = 17.7 \quad ; \quad n_{M_*} = nn_\Delta^{-1} n_D^{-1} = 21$$
$$n_{\rho_*} = 1 \quad ; \quad n_{G_*} = nn_D^{-1} = 20, \tag{49}$$

which indicates some serious scale effects, particularly because both M_* and G_* in the model are an order too small.

7.4. Model distortion

In the above discussion, geometric distortion of the model is presented as a parameter which can be introduced at will by the modeler. For mobile bed coastal models, however, there are some severe restrictions. Ideally, particularly if the model involves structures, the model should be undistorted. On the other hand, space limitation may require a certain distortion. But mobile bed models have what could be termed as a *natural distortion*. The beach in a model is formed by the waves and currents and will take on a profile which is a function of these conditions and the composition of the beach. For example, a sand model beach is usually steeper than its sandy beach prototype.

Natural distortion may be defined as

$$N_n = \frac{n_x}{n} = \frac{x_P}{x_M} \frac{z_M}{z_P} = \frac{m_M}{m_P} = n_m^{-1}. \tag{50}$$

To determine N_n it is necessary to perform preliminary model tests. From model beach slopes in these tests, it is possible with Eq. (50) to determine N_n. Kamphuis (1995) shows that these tests should be three-dimensional, but that two-dimensional (flume) tests are a reasonable approximation. Since a number of such preliminary tests would be required using different model grain sizes, it is useful, for model planning purposes, to be able to calculate natural distortion approximately from known equations. Kamphuis (1995) shows that a carefully modeled beach approximately follows the well-known prototype expression developed by Bruun (1954) and Dean (1977):

$$d = Bx^{2/3} \tag{51}$$

where B is a rather complicated function of grain size. Beach slope and the beach slope scale may now be derived from Eq. (51). One simplification which is useful in model design will now be demonstrated. Figure 1 shows the curve for B vs. D based on the data published by Moore (1982), Dean (1983), or CUR (1990). Although the complete relationship is rather complicated, for physical modeling, a simple power relationship between B and D may be developed which specifically only includes the required model and prototype grain sizes. Two such example relationships are

$$B = 0.27\, D^{0.64} \qquad 0.1 \leq D \leq 0.25$$
$$B = 0.23\, D^{0.23} \qquad 2 \leq D \leq 100. \tag{52}$$

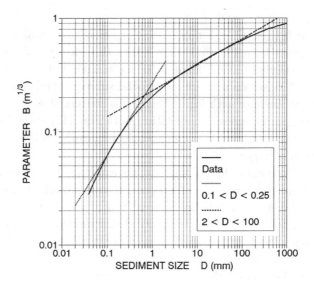

Fig. 1. Determining a simple power relationship for beach profile parameter B.

Here, the units for B and D are $m^{1/3}$ and mm respectively. Such simple power relationships can be readily converted to model scales. For the earlier example of the sand model ($D_M = 0.1$ mm) of a sand prototype ($D_p = 0.25$ mm), this means

$$m = \frac{d}{x} = Bx^{-1/3} \quad \rightarrow \quad n_m = \frac{n_B}{N^{1/3}n^{1/3}} = \frac{n_D^{0.64}}{N^{1/3}n^{1/3}} \tag{53}$$

From Eqs. (50) and (53), the natural distortion may be derived as

$$N_n = n_m^{-1} = \frac{N_n^{1/3} n^{1/3}}{n_B} \quad \rightarrow \quad N_n = \frac{n^{1/2}}{n_B^{3/2}} = \frac{n^{1/2}}{n_D^{0.96}}. \tag{54}$$

For our example in which $n = 50$ and $n_D = 2.5$, $N_n = 2.9$.

Other recipes may also be found in the literature to serve as a first approximation. They are based on eroding beach results or dune model results in which the eroded material is deposited offshore. Hughes (1993) summarizes these test results to find

$$N_n = \left(\frac{n}{n_\omega^2}\right)^\theta \quad \text{where} \quad 0.25 \leq \theta \leq 0.50. \tag{55}$$

Hallermeier (1981) and the Shore Protection Manual (1984) find that $\omega \sim \Delta^{0.7} D^{1.1}$ for $0.13 < D < 1$ mm. Using $n_\omega \approx n_\Delta^{0.7} n_D^{1.1}$ makes Eq. (55) only slightly different from Eq. (54) for $\theta = 0.5$. But for $\theta = 0.28$, resulting from the work of Vellinga (1986), for $n = 50$, $n_\Delta = 0.96$, and $n_D = 2.5$, N_n is 1.7. Equation (55), based on onshore erosion and subsequent offshore deposition, is also clearly at variance with Eq. (47) which is concerned with only deposition. Only an undistorted model would satisfy both Eqs. (47) and (55). This would necessitate making $n_\omega = n^{1/2}$, which would be possible if the prototype grain size is relatively large.

To model gravel or shingle beaches, for example, $D_p = 100$ mm and $n = n_D = 50$ would result in $D_M = 2$ mm and the second relationship of Eq. (52) applies. Equation (54) results in $N_n = n^{1/2}/n_D^{0.36} = 1.7$. For these grain sizes, Hallermeier (1981) says that $\omega \sim (\Delta D)^{1/2}$ and therefore Eq. (55) will always yield an almost undistorted model. However, neither Eq. (51) nor Eq. (55) was developed for shingle beaches and hence the more complex relationships for shingle and gravel beaches that may be found in the literature, for example CIRIA/CUR (1991), should perhaps be used.

In any case, computations provide only rough guidelines about the natural distortion of the model and the preliminary model tests should certainly be performed to determine natural distortion accurately.

8. Implications for Model Design

For a wave model with sediment transport and erosion close to shore and deposition further offshore, Eqs. (35), (36), (47), and (48) as well as model distortion must be considered simultaneously. It is not difficult to see that while these equations are helpful for understanding the processes involved and for determining how to build models, it is not possible to satisfy all of the scaling criteria developed above, even for a relatively simple coastal model.

For many researchers, particularly those involved with numerical modeling and field studies only, this all too often means rejecting this truly incredible tool out of hand. First, it is simply stated that one must look to field experiments for "real" data. Second, because physical models are fraught with scale effects, it is said that we must turn to numerical modeling. The appealing graphics produced by numerical models will also help to convince the client that the new methods are better. With respect to the graphics, Abott (1991) however warns that "the images need not necessarily correspond to physical reality".

The researcher familiar with physical modeling, however, wants to make use of the very real and important qualitative impressions provided by a physical model. Although these images also need not correspond to reality, at least they are based on a degree of physical similarity so that many of the complex processes, their interactions, and the complicated boundary conditions are all reasonably modeled. The real difference between a physical and numerical model becomes evident in this qualitative phase of the modeling process. The results of a numerical model simply *echo* the concepts and ideas fed into the computer by the modeler. They do not go beyond the input equations and coefficients. A physical model, on the other hand, *adds* to the modeler's input. The actual physical similarity of the model and the prototype fills in many details that the modeler did not or could not consider in the original model design.

The physical modeler also knows that a model study cannot be quantitatively successful unless the modeler understands the shortcomings of the model and works around them as best possible. Only a modeler who understands which parameters are important and what are the scale effects resulting from ignoring some of the scaling laws, a modeler who goes through detailed reasoning similar to that in the above sections will be successful in obtaining good quantitative results — results which neither numerical modeling nor field experimentation can provide. Kamphuis (1991) argues that the hydraulic model is the pivotal study tool that can link the other two methods together so that all three become useful. Le Méhauté (1990) says: "A return to experimental techniques, which may have declined during the last two decades because of the successful development of mathematical modeling appears not only desirable but inevitable. The need for experimentation will actually be dictated by (the gaps in) mathematical modeling."

Clearly hydraulic modeling is not simply a matter of applying some *recipes* based on equations, dimensional analysis, or both. Dimensional analysis can best be used to organize and understand the problem and to set up a scaling framework.[c] Any available equations can then be applied to incorporate scaling for what is known theoretically or empirically. The modeler must then carefully sift through the resulting scale effects and build a model that best simulates those aspects of the prototype which are of greatest interest. This may mean that several models are necessary to study different aspects of one prototype.

To understand scale effects better, a *scale series* (models of the same prototype, but built at different scales) may be used. Scale series testing is also very useful if inadequate prototype information is available. Many prototype results may then be inferred from the model studies by extrapolation. Care should be taken with this method, however. Ackers (1990) is correct when he says: "The reliability of scale models cannot be judged by theoretical exercises or from laboratory scale tests alone. Well-documented measurements from prototype behavior are the only proof of success."

It should also be clear that the *long-term design model* suffers most from scale and laboratory effects. Since a particular prototype must be modeled, the modeler is left with few options but to build the best possible model, calibrate it, and live with all the scale and laboratory effects. As a result, the discussion in the above sections on scaling becomes mainly academic and only relevant in the broadest sense for planning such models. Technically, such models should only be useful to give qualitative results. Le Méhauté (1990) states, however, that scale models remain the "best analog computers". The value of the qualitative results must not be underestimated. They describe and often redefine the problem and give many indications toward possible solutions. On the other hand, these qualitative results cannot be simply assumed to be quantitative. Any quantitative results derived from such models can normally be directly attributed to the understanding, experience, and ingenuity of the modeler.

Historically, the long-term design model was the most common type of model and because of its requirements for large laboratory space, it gave birth to the great laboratory facilities that were built earlier in this century.

---

[c]In the same manner, dimensional analysis could also be used to plan, organize, and understand field experiments and exercises in numerical modeling.

Often such models are still a good shortcut to results when answers are needed and the intermediate steps (equations and interactions of various parts of the model) are not clearly understood. But because of their limitations, the use of such models has waned and as a result the great modeling laboratories have experienced difficult times since the 1970s.

The *short-term design* model offers more options. Input prototype conditions are better defined and simpler which means that laboratory effects are usually smaller. The sections of the prototype tested are usually smaller than for the long-term design model. This means smaller model scales (Eq. (5)) become economically possible and this has given rise to the popularity of large wave flumes to reduce scale and laboratory effects. The discussion on scaling becomes much more relevant for these models. The cost of such a model also does not depend on the time scales to the same extent as for the long term design model which means that model tests are short and therefore can readily be repeated and scale series tests can be contemplated.

Because the *process model* is really an abstraction of prototype conditions, it offers the modeler even more alternatives. It will be set up specifically to minimize laboratory and scale effects and thus the process model will yield the best quantitative results. The earlier discussion on scale selection and scale effects is now obviously very relevant and the details presented must be clearly understood by the modeler for such models to be successful.

The trend for more accurate and quantitative answers to more precise questions requires, among other things, that the model results can be trusted without putting undue faith in the modeler. Because of the scaling and laboratory limitations, physical modeling has therefore the tendency to move away from the long-term design model toward the process model. This trend will continue. It means proper understanding of the details of model scaling as demonstrated above will become ever more important and relevant. One indication of this is the recent publication of four large volumes which deal primarily with scaling of physical models (Martins, 1989; Shen, 1990; Hughes, 1993; and Chakrabarti, 1994).

Along with the move toward more precise process modeling comes the tendency to reduce laboratory effects as much as possible by improving laboratory equipment and methodology. For a long-term design model, the many simplifying assumptions meant that the actual model input conditions could be quite crude. Since, for a process model, the input conditions can be quite accurately specified, it is reasonable that more care is taken in the model to simulate the input conditions correctly. This has resulted in rapid improvements to modeling equipment. For example, in only a few years, wave generation has gone from paddle-generated regular waves, through long-crested, irregular waves to directional seas. It is now also possible to suppress unwanted long-wave activity and Dalrymple (1989) talks about "designer" waves in which the computer program driving the directional wave generator takes into account the reflections off the side walls of the basin to produce a prescribed wave exactly at the structure to be tested.

Process model results should not be expected to provide direct solutions to practical problems; they must be seen as building blocks which may be used as steps toward a solution of a problem. The mortar that will eventually hold these building blocks together is computer calculations. The complete modeling task produces a *composite model* consisting of three distinct phases — a physical modeling phase, the analysis of the physical model results, and a computation phase. The computational model could be a complete numerical model in which case the physical process models simply provide appropriate coefficients and transfer functions. It is more likely, however, for engineering studies that need to provide useful answers within a limited time and budget for situations with complicated boundary conditions, that the computational model will be a relatively simple statistical summation of a number of carefully determined process modeling results or of the relatively simple empirical relationships derived from such model results. Nonlinear interactions between the various building block units need to be carefully investigated from the process model results.

As an example, consider long-term scour near tidal inlet structures. This process is impossible to study effectively with a long-term design model. Short-term design models could perhaps provide answers with respect to limiting scour conditions. It is possible, however, to conduct a large number of relatively similar, easy-to-set-up process studies, each simulating scour for individual combinations of one current direction and velocity, one wave height, one wave period, and one incident wave angle. Nonlinear interaction can be derived from the time evolution of the scour holes. Ideally, empirical relationships can be derived from the individual model studies to be combined in computer calculations which reflect wave and tidal current statistics to produce short- and long-term erosion rates and volumes, limiting states, etc. To understand scale effects better, scale series tests can also be done. This example of composite modeling is shown in more detail in Figs. 2 and 3.

OBJECTIVE: To determine scour near inlet structures.

EQUATION: $S = f(H, T, \alpha, V, D, Q, n, t)$

DEFINITIONS:
    VARIABLES:
        H - Wave Height
        T - Wave Period
        α - Incident Wave Angle
        V - Current Velocity
    CONDITIONS:
        D - Direction of Tidal Current (Ebb/Flood)
        Q - Littoral Transport Direction (North/South)
        n - Model Scale (15, 30 or 50)

BASIC TEST SERIES:

a) BASE TEST: (Most important combination of variabless and conditions)
    α_B, H_B, T_B, V_B, D=Ebb, Q=North, n=30

b) CHANGING VARIABLES: ($^+$ = larger, $^-$ = smaller)

V-1	α_B, H_B^+, T_B, V_B, D=Ebb, Q=North, n=30
V-2	α_B, H_B^-, T_B, V_B, D=Ebb, Q=North, n=30
V-3	α_B, H_B, T_B^+, V_B, D=Ebb, Q=North, n=30
V-4	α_B, H_B, T_B^-, V_B, D=Ebb, Q=North, n=30
V-5	α_B^+, H_B, T_B, V_B, D=Ebb, Q=North, n=30
V-6	α_B^-, H_B, T_B, V_B, D=Ebb, Q=North, n=30
V-7	α_B, H_B, T_B, V_B^+, D=Ebb, Q=North, n=30
V-8	α_B, H_B, T_B, V_B^-, D=Ebb, Q=North, n=30

c) CHANGING CONDITIONS:

C-1	α_B, H_B, T_B, V_B, D=Flood, Q=North, n=30
C-2	α_B, H_B, T_B, V_B, D=Ebb, Q=South, n=30
C-3	α_B, H_B, T_B, V_B, D=Flood, Q=South, n=30
C-4	α_B, H_B, T_B, V_B, D=Ebb, Q=North, n=15
C-5	α_B, H_B, T_B, V_B, D=Ebb, Q=North, n=50

d) NOTES:
- Complete test matrix comprises 3x3x3x3x2x2x3 = 972 tests.
- Basic test series consists of: 1+8+5 = 14 tests.
- Basic test series will help identify additional tests, needed to cover especially important combinations, non-linear interactions, test repeatability, etc.
- For change of conditions, the base variables α_B, H_B, T_B and V_B may not be most appropriate.

Fig. 2. Process modeling of scour at inlet structures.

Figure 2 describes the basic process modeling tests that would be needed to start such a project. Figure 3 is a schematic which outlines the composite modeling steps.

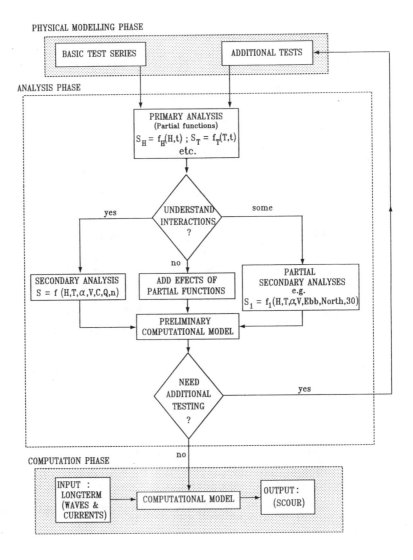

Fig. 3. Schematic for composite modeling of scour at inlet structures.

Composite modeling has many distinct advantages over either pure physical modeling or pure numerical modeling. Because scale and laboratory effects are limited in process models, the main drawback of physical modeling has been reduced. Because physical modeling results are included in any numerical calculations, the output goes far beyond simply echoing input equations. Because both modeling concepts are combined, the method is immediately useful for problems that cannot be solved by either, by drawing on the strengths of both methods.

Several aspects of composite modeling also make it economically attractive. First, the physical models are *simpler* (with respect to the scaling relationships), *less expensive*, and *easy to understand*. They are repeatable and because the tests are very similar to each other, the experience gained with the first studies is immediately used in later similar studies resulting in high efficiency.

Second, the physical model results used in composite modeling are *generic* which means they are not very site specific and may be used to solve many similar problems for totally different layouts and locations. One could visualize, in time, complete libraries of such physical process modeling results which can be combined computationally to solve many different problems, greatly reducing the number of new model tests actually required to solve any particular problem.

A third interesting aspect is that model calibration takes place within the computation phase. This permits calibration, verification, and as many "what-if" scenarios as the modeler is willing to investigate at a very *low cost*.

Finally, the physical modeling and the computation phases of a composite model study need not be carried out by the same organization. For example, an informed *client* could also do the scenario computations once the laboratory has provided the building blocks.

9. Conclusions

Physical models are a unique tool. Their scales are based on the use of dimensional analysis and known equations but the model itself transcends both these representations of the prototype. It provides much more detail than dimensional analysis. At the same time it represents some nonlinear and complex processes and boundary conditions that cannot be described by equations.

Because of the close physical similarity to the prototype, a physical model immediately provides valuable qualitative results and impressions.

Physical models are plagued by scale and laboratory effects because the model can never accomplish perfect simulation. This means common sense, intuition, and experience of the modeler as well as the modeler's knowledge of scale and laboratory effects are needed to convert the initial qualitative results into quantitative answers.

For coastal models, lightweight modeling materials should be avoided. Geometric distortion should also be avoided, if diffraction and erosion and settling must all be modeled correctly. However, the model has a mind of its own, introducing a natural distortion, which must be taken into account by the modeler.

The requirements for more scientific results in the future will continue a trend in physical modeling away from the overall long-term design models of immediately practical situations toward more abstract process models. Process models are accurate analogs of relatively simple prototype conditions. Accuracy is achieved by minimizing scale and laboratory effects and introducing relatively simple boundary conditions more correctly. The use of process models will require a more thorough understanding of model scaling and scale effects than was needed in the past.

To obtain practical answers from the rather abstract process models, it will be necessary to link many individual process model results by computer calculations (composite modeling). This effective modeling technique combines the strengths of both physical and numerical modeling. It is economically attractive, particularly because the calibration and "what-if" scenario work is done in the relatively inexpensive computation phase.

Acknowledgements

This review article is based on the author's experience with physical models at the Queen's University Coastal Engineering Research Laboratory and elsewhere. There are too many sponsors for such models to acknowledge individually, but the funding for the basic research on hydraulic modeling of coasts and for equipment provided since 1972 by the Natural Sciences and Engineering Research Council of Canada is gratefully acknowledged. Thanks are also extended to Dr. G. Mogridge of the National Research Council of Canada, who provided the information for the Ocean Ranger example and to Dr. S. Hughes of the Coastal Engineering Research Center of the US Army Corps of Engineers who reviewed the draft and provided helpful comments.

References

Abbott, M. B. (1991). Numerical modeling for coastal and ocean engineering. *Handbook of Coastal and Ocean Engineering*, Vol. 2, ed. J. Herbich, Gulf Publishing. Ch. 22, 1067–1124.

Ackers, P. (1990). Dimensional analysis, dynamic similarity, process functions, empirical equations and experience — how useful are they? *Movable Bed Physical Models*, ed. H. W. Shen, Kluwer Academic Publishers. 23–30.

Benson, M. A. (1965). Spurious correlation in hydraulics and hydrology. *J. of Hydraulics. ASCE.* **91**: 35–42.

Bruun, P. (1954). Coast erosion and development of beach profiles. Beach Erosion Board, US Army Corps of Engineers, Vicksburg, USA, Technical Memo No. 44.

Chakrabarti, S. K. (1994). *Offshore Structure Modeling, Advanced Series on Ocean Engineering*, Vol. 9. World Scientific Publishing. 470.

CIRIA/CUR (1991). *Manual on the Use of Rock in Coastal and Shoreline Engineering*. CIRIA/CUR Rep 154. Balkema. 607.

CUR (1990). *Manual on Artificial Beach Nourishment.* CUR Rep 130, Balkema. 193.

Dalrymple, R. A. (1989). Physical modeling of littoral processes. *Recent Advances in Physical Modeling*, ed. R. Martins. Kluwer Academic Publishers. 567–588.

Dean, R. G. (1977). Equilibrium beach profiles: US Atlantic and Gulf Coasts. University of Delaware, Newark, USA, Coastal Engineering Report No. 12.

Dean, R. G. (1983). Principles of beach nourishment. *Handbook of Coastal Processes and Erosion*, ed. P. Komar. CRC Press. 217–232.

Fredsøe, J. and R. Deigaard (1992). *Mechanics of Coastal Sediment Transport. Advanced Series on Ocean Engineering*, Vol. 3. World Scientific Publishing. 367.

Hallermeier, R. J. (1981). Terminal settling velocity of commonly occurring sand grains. *Sedimentology* **28** (6): 859–865.

Hanson, H. and N. C. Kraus (1989). Genesis: Generalized model for simulating shoreline change. Coastal Engineering Research Center, US Army Corps of Engineers, Vickburg, USA, Report 89-19.

Hughes, S. (1993). *Physical Models and Laboratory Techniques in Coastal Engineering, Advanced Series on Ocean Engineering*, Vol. 7. World Scientific Publishing. 567.

Ivicsics, L. (1980). *Hydraulic Models*. Water Resources Publications. 310.

Kamphuis, J. W. (1974). Determination of sand grain roughness for fixed beds. *J. Hyd. Res. IAHR* **12** (2): 193–207.

Kamphuis, J. W. (1975). The coastal mobile bed model. Queen's University, Kingston, Canada, CE Report No. 75. 114.

Kamphuis, J. W. (1985). On understanding scale effect in coastal mobile bed models. *Physical Modeling in Coastal Engineering*, ed. E. Dalrymple. Balkema. 141–162.

Kamphuis, J. W. (1988). On bedform geometry and friction. *Proc. Workshop on Roughness and Friction.* National Research Company of Canada. 19–35.

Kamphuis, J. W. (1991). Physical modeling. *Handbook of Coastal and Ocean Engineering*, ed. J. Herbich. Gulf Publishing. Ch. 21, 1049–1066.

Kamphuis, J. W. (1995), Comparison of two-dimensional and three-dimensional beach profiles. *J. Wtrwy., Harb. and Coast. Eng. ASCE* **121** (3): 155–161.

Kamphuis, J. W. and M. S. Yalin (1970). Theory of dimensions and spurious correlation. *J. Hyd. Res. IAHR* **9**: 249–266.

Kriebel, D. L., W. R. Dally, and R. G. Dean (1986). Undistorted Froude model for surf zone sediment transport. *Proc. 20th Int. Conf. Coastal Engineering. ASCE.* 1296–1310.

Langhaar, H. L. (1951). *Dimensional Analysis and Theory of Models*. John Wiley and Sons. 165.

Le Méhauté, B. (1990). Similitude. *Ocean Engineering Science, The Sea*, Vol. 9. John Wiley and Sons. 955–980.

Martins, R. (1989). *Recent Advances in Physical Modeling*, ed. R. Martins. Kluwer Academic Publishers. 620.

Mogridge, G. R. (1985), Hydrodynamic model tests of the semi-submersible "Ocean Ranger". National Research Company of Canada, Report TR-HY-007. 227.

Mogridge, G. R. and J. W. Kamphuis (1972). Experiments on ripple formation under wave action. *Proc. 13th Int Conf. Coastal Engineering*, Vancouver, Canada. ASCE. 1123–1142.

Mogridge, G. R., B. D. Pratte, and W. W. Jamieson (1986). Hydrodynamic model study of the semi-submersible "Ocean Ranger". *Proc. 5th Int. Offshore Mechanics and Arctic Engineering Symp.*, Vol. III. OMAE. 1–8.

Moore, B. (1982). Beach profile evolution in response to changes in water level and wave height. M.Sc. Thesis, University of Delaware, Newark, USA.

Noda, E. K. (1972). Equilibrium beach profile scale model relationship. *J. Wtrwy., Harb. and Coast. Eng. ASCE* **98**: 511–528.

Schuring, D. J. (1977). *Scale Models in Engineering*. Pergamon Press. 285

Sharp, J. J. (1981). *Hydraulic Modeling*. Butterworths. 240.

Shen, H. W. (1990). *Movable Bed Physical Models*, ed. H. W. Shen. Kluwer Academic Publishers. 1–12.

Shen, H. W. (1990a). Introductory remark. *Movable Bed Physical Models*, ed. H. W. Shen. Kluwer Academic Publishers. 1–12.

US Army Corps of Engineers (1984). *Shore Protection Manual*. Coastal Engineering Research Center, US Army Corps of Engineers, Vicksburg, USA. 1230.

Vellinga, P. (1986). Beach and dune erosion during storm surges. Delft Hydraulics, Delft, The Netherlands, Communication No. 372.

Yalin, M. S. (1971). *Theory of Hydraulic Models*. MacMillan Press. 266.

Symbols

A	Dependent variable (dimensional)
a	Independent variable (dimensional)
a_B	Wave orbital amplitude at the bottom
B	Beach shape parameter
b (subscript)	Breaking
c (subscript)	Current
D	Grain diameter
d	Depth
f	Function (dimensional)
f (subscript)	Flow
G_*	Geometric link between waves and sediment — (a_B/D)
g	Gravitational acceleration
H	Wave height
k	Bottom roughness
L	Wave length
$[L]$	Length (base unit)
M_*	Mobility number — $(v_*^2/\Delta D)$
$[M]$	Mass (base unit)
M (subscript)	Model
m	Beach slope
N	Geometric model distortion
N (when subscripted)	Subscript denotes type of distortion
n	General (vertical) model scale
n (when subscripted)	Subscript denotes scale
n (subscript)	Natural
P (subscript)	Prototype
p	pressure
R_*	Grain size Reynolds number — $(v_* D/\nu)$
S	Radiation stress
T	Wave period
$[T]$	Time (base unit)
t	Time
U	Depth averaged velocity component in the x direction
u	Horizontal component of particle velocity in the x direction
u'	Turbulent velocity fluctuation in the x direction
u_w	Wave orbital velocity component in the x direction
V	Depth averaged velocity component in the y direction
v	Horizontal component of particle velocity in the y direction
v'	Turbulent velocity fluctuation in the y direction
v_*	Shear velocity — $(\tau/\rho)^{1/2}$
w	Vertical component of particle velocity in the z direction
w (subscript)	Waves

w'	Turbulent velocity fluctuation in the z direction
X	Independent variable (dimensionless)
x	Horizontal coordinate direction offshore
y	Horizontal coordinate direction alongshore
z	Vertical coordinate direction down from still water
α	Incident wave angle
β	Exponent
Δ	Relative underwater sediment density — $(\rho_s - \rho)/\rho$
δ	Bedform height
η	Wave set-up
θ	Exponent in natural distortion expression
μ	Dynamic viscosity of the fluid
ν	Kinematic viscosity of the fluid
Π	Dependent variable (dimensionless)
ρ	Density of the fluid
ρ_s	Density of the sediment
ρ_*	Relative sediment density — ρ_s/ρ
τ	Shear stress
ϕ	Function (dimensionless)
ω	Fall velocity

MATHEMATICAL MODELING OF MESO-TIDAL BARRIER ISLAND COASTS
Part I: EMPIRICAL AND SEMI-EMPIRICAL MODELS

HUIB J. DE VRIEND

Starting from a typology of phenomena and models for large meso-tidal inlets, various forms of empirical and semi-empirical models are outlined and discussed. They range from geostatistical models, via empirical equilibrium-state relationships, to semi-empirical models of various degrees of resolution for the basin and the outer coast. Examples of applications to "mixed-energy tide-dominant" inlets are shown and needs for further research are identified.

Process-based simulation, a very important approach in tidal-inlet modeling, is discussed in the chapter.

1. Introduction

A large part of the world's coastline consists of more or less elongated barrier islands. They are separated from the mainland by a shallow tidal basin and from each other by relatively narrow inlets to that basin. For various reasons, these barrier island coasts are important issues in coastal zone management.

Barrier islands tend to be morphologically quite active, hence they are difficult to control by engineering measures. Yet, many of these islands are inhabited, and the houses and properties often require protection. This is a continuous source of problems at the engineering level.

The outer deltas in front of the inlets store large amounts of well-sorted sand, which is attractive to be mined for all sorts of purposes (for instance, beach nourishment). As the delta is part of a coherent morphological system, however, this sand mining may have detrimental effects on the other parts of the system (viz. the adjacent island coasts and the basin).

The sheltered borders of estuaries and lagoons are attractive places for economical and recreational activities related to the sea. This explains why many inlets are important navigation channels. As inlet channels tend to be morphologically even more active than barrier islands, it is tempting to try and stabilize them with engineering works like jetties and revetments. The connection of such works with the mobile natural system is rather problematic and costly.

Many tidal lagoons, especially the intertidal zones, are of paramount importance to the ecosystem as feeding and breeding grounds for many species throughout the food chain. Thus, these areas have a high nature-preservation status. At the same time, the system is quite vulnerable because the intertidal area is very sensitive to changes in the physical system (e.g., accelerated mean sea-level rise).

Since tidal-inlet systems respond to the mean sea level, the tide, and the wave conditions, they are not only threatened by sea-level rise and subsidence. Also, changes in the tidal regime (amplitude, asymmetry) and the wave climate (mean wave height and direction, storm climate, chronology of events) can have a major impact.

The ability to understand and predict the morphological behavior of barrier island coasts, including inlets and basins, is therefore an important issue in coastal zone management. Mathematical models are particularly attractive here because of their flexibility and because the complexity of the system makes physical scale modeling very difficult, especially when it concerns long-term morphological evolution.

In recent years, tidal inlets have come within the reach of mathematical modelers. It has become possible to model many of the complex physical processes (waves, currents, sediment transport, morphological changes) which take place in the vicinity of an inlet. Such models enable researchers to develop and test hypotheses on how an inlet system works, and to identify the most important knowledge gaps. They also enable practitioners to analyze the response of a given inlet system, for instance, to proposed engineering measures.

In spite of the computer technology available at present, these process-based simulation models are as yet unable to cover time spans which are much larger than the principal inherent morphological time scales. Much of the empirical knowledge on the equilibrium state of tidal inlets, for instance, cannot yet be reproduced by these models. This has boosted the formulation of semi-empirical long-term models which are based on elementary physics, but include the empirical knowledge.

As it is hardly possible to cover the entire range of coastal inlets, this chapter focuses on what Hayes (1979) calls *mixed-energy tide-dominant inlets*. This is the type of inlets which is common along the Northwest European coast for instance. The tidal range is typically a few meters and the yearly average significant wave height is typically one meter. In their natural state, these inlets are rather wide, with a large dynamic outer delta and drumstick-shaped barrier islands.

After a brief outline of the practically relevant phenomena, which inlet models should be able to cover, the various model types are classified. Except for the process-based simulation models, which are treated in the next chapter, each of the classes is described, and its potentials, shortcomings, and perspectives are discussed.

2. Phenomena to be Modeled

In the present state of development, it is not realistic to claim that one model type can provide the necessary morphological information for all possible practical problems. In other words, the all-purpose model for tidal inlet morphodynamics does not exist and is not likely to emerge in the near future. This means that the present models are linked to classes of phenomena.

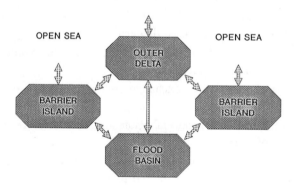

Fig. 1. Large-scale tidal inlet system.

Barrier islands, outer deltas, inlet channels, and flood basins are all elements of one coherent morphodynamic system (Fig. 1). The nature of this "tidal inlet system", however simple at first sight, is very complex. It is forced by a partly unpredictable input signal (waves, storm surges), and its behavior involves a wide range of space and time scales. Part of this behavior is imposed

118 H. J. de Vriend

by variations of the forcing signal, part is inherent to the system. An example of forced behavior is the 18.6-yearly variation of the inlet geometry due to the nodal cycle of the tide (Oost et al., 1993). An example of inherent behavior is the formation and migration of channel and shoal patterns in the flood basin and on the delta fringe (for instance, see Ehlers, 1988).

The complexity and the scale range of this inherent behavior make tidal inlets particularly difficult to model. Models which perform well in describing small-scale phenomena are not necessarily suited for long-term predictions, even if there is ample computer time available. The phenomena which a model is able to describe are therefore classified relative to the inherent scale of the corresponding morphodynamic process. The latter represents both the space and the time scale because these are assumed to be coupled: Larger-scale phenomena will proceed more slowly than small-scale ones (cf. De Vriend, 1991; Fenster et al., 1993). This leads to the following classification:

(1) Microscale (*process-scale*) phenomena take place at an essentially smaller scale than the corresponding morphodynamic behavior. It concerns primarily the constituent processes (waves, currents, and sediment transport). Examples of relevant information are:
    - wave conditions in channels, on shoals, and near structures (relevance: navigation, habitat, coastal defence);
    - current speed in channels, on shoals, and near structures (relevance: navigation, habitat, local scour);
    - patterns of water mass exchange (relevance: water quality, import of nutrients, larvae, etc.);
    - suspended sediment concentration and turbidity (relevance: habitat, mobility of channels, and shoals);
    - sediment transport rate (relevance: mobility of channels and shoals; import of adhering nutrients and pollutants).

(2) Mesoscale (*dynamic scale*) phenomena concern the "primary" morphodynamic behavior, due to the interaction of the constituent processes and the bed topography. The scales involved are of the order of magnitude of those inherent to this interaction, which is generally much larger than those of the constituent processes. Examples of relevant information are:
    - bedform (ripples, dunes, sandwaves) dimensions, propagation, and patterns (relevance: influence on current and transport patterns, depth for navigation, risk of denudation of pipelines);

- shoaling and migration rates of natural channels (relevance: navigation, coastal defence);
- reside time of bottom pollutants (relevance: water quality, habitat);
- morphological changes around man-made works (relevance: stability of structures, (un)burial of pipelines, shoaling of dredged navigation channels, filling up of sand-mining pits);
- response to human interference, such as land reclamation, persistent sand mining, repeated channel dredging, etc. (relevance: to avoid detrimental effects elsewhere in the system).

(3) Macroscale (*trend-scale*) phenomena concern slow trends at scales much larger than those of the "primary" morphodynamic behavior. These trends can be due to secular effects in the system's inherent behavior, or to gradual changes in the extrinsic forcing or the system parameters. The cyclic migration of channels in a wide inlet (Bakker and Joustra, 1970; Oost and De Haas, 1993; Oost, 1995) may be considered as a secular effect. Gradual changes in the extrinsic forcing may be due to mean sea-level rise, subsidence, tidal range evolution, etc. Erodibility is an example of a system-parameter which is affected by changes in the cohesive sediment content, the intensity of mussel fishing, and the marshland vegetation (due to pollution, for instance). Examples of relevant phenomena are:
- evolution of channel/shoal patterns in the basin and on the delta (relevance: habitat and ecosystem, coastal defence at barrier islands, navigability, stability of pipelines);
- evolution of the intertidal zone (aspects: area, level, duration of flooding and drying) (relevance: habitat and ecosystem, (indirectly) coastal defence at barrier islands);
- inlet instability (relevance: coastal defence, ecosystem, navigation).

The above classification is relative, so the scale classes cannot be expressed in absolute terms of meters or years. Figure 2 illustrates that the three classes exist for a wide range of phenomena at different scales.

3. Classification of Mathematical Models

The statement that the all-purpose model does not exist is substantiated by the wide variety of tidal inlet models which have been proposed. The above classification of phenomena can be used to structure this set.

120 H. J. de Vriend

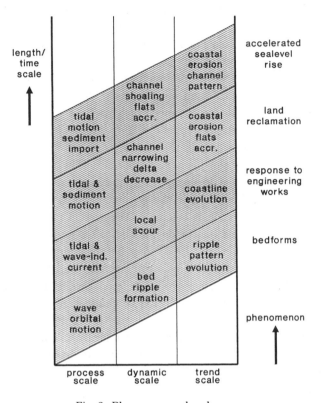

Fig. 2. Phenomena and scales.

Another axis of classification is the model approach. In a broad outline, the following model types can be distinguished:

- *Data-based models*: making use of measured data only and aiming directly at the phenomena to be described.
- *Empirical relationships*: also observation-based, but establishing a relationship between input and output of the system.
- *Process-based models*: describe waves, currents, sediment transport, and bed level changes via a set of mathematical equations based on first physical principles (conservation of mass, momentum, energy, etc.).
- *Formally integrated, long-term models*: derived from a process-based model by formal integration over time (and space), with empirical or parametric closure relationships.

- *Semi-empirical, long-term models*: describe the dynamic interaction between large elements of the system, using empirical relationships to represent the effects of smaller-scale processes.

Figure 3 shows that these two axes of classification are not entirely orthogonal. Henceforth, the various model types will therefore be distinguished by approach only.

	process scale	dynamic scale	trend scale
process-based models	X	X	
formally integrated models	x	x	x?
semi-emp. models		x	X
data-based models			X
empirical relationships			X

Fig. 3. Mathematical models and phenomena.

4. Data-Based Models

Geostatistical models belong to the class of data-based models. The simplest form of geostatistical modeling is the extrapolation in time of a certain parameter of the system's state (for instance the minimum depth of a certain channel) via a linear regression analysis of observed values. The model essentially consists of the coefficients which result from the regression analysis, together with the assumption that the parameter to be predicted will continue its past evolution. The underlying assumption is that the processes which determine the trend in this evolution will remain constant.

Fenster *et al.* (1993) decribe a more sophisticated version for the multiscale nonlinear system of an uninterrupted coast. Although inlet systems are essentially more complicated, the technique should be transferable to tidal inlets. A necessary condition for validity as a prediction tool is that the processes which lead to the described behavior will remain unaltered in the future.

Another form of data-based modeling is "translation" of observations at one site to predictions at another. This approach may be applicable if the system is to undergo changes (due to engineering works, for instance) which have taken place before in a similar, well-monitored system. The underlying assumption is that the two systems will exhibit a similar response to this interference.

Empirical Orthogonal Function (EOF) analysis (Preisendorfer and Mobley, 1988) is a form of data-based modeling which separates various modes of variation in space and time. The weighted sum of products of these modes constitutes the complete data set, and for each of these products the content of measured information can be assessed. Thus, EOF modes can be ordered according to their content of information on the observed behavior. In the case of uninterrupted coasts, these series used to converge rapidly, which means that the behavior can largely be mapped onto a few EOF modes (cf. Medina *et al.*, 1992; Wijnberg and Wolf, 1994; also see De Vriend *et al.*, 1993a). As yet, applications to tidal inlet systems are scarce, so it is unclear whether this favorable property also applies there.

In general, EOF analysis allows for a certain degree of physical interpretation per component. Hence, extrapolation into the future can be done per component based on physical considerations. At any future time, the state of the system can be composed of various components.

Data-based models of tidal inlets typically concern macroscale phenomena, such as the historical or geological evolution of the system as a whole. Their concept ("let nature tell what it is like") may seem attractive for complex systems because it seems to evade the necessity to analyze the system's behavior. It is exactly this complexity, however, which rules out such a "black-box" approach. In simple cases it is relatively easy to predict whether or not the relevant conditions and processes are likely to remain unaltered. The system's behavior is transparent and it is determined by one or a few predominant mechanisms. The relevant input and process parameters are not many, which makes it easier to predict whether the system's behavior is likely to change in the future.

In a system as complex as a tidal inlet, however, there are many input and process parameters to take into account, and the system's behavior is difficult to analyze. Hence, it is difficult to asses which mechanisms determine which aspect of the system's evolution. Utilizing data-based models in such a situation therefore requires an equally thorough insight into the mechanisms at work as any physics-based model.

5. Empirical Models

5.1. *Equilibrium-state relationships*

O'Brien (1931, 1969) was the first to suggest a linear relationship between the cross-sectional area of an inlet and the tidal prism, defined as the total amount of water which moves in and out the inlet during one tide,

$$A_c = c_A P \tag{1}$$

in which

A_c = cross-sectional area of the inlet channel below MSL [m^2],
c_A = coefficient of proportionality; Eysink (1990) suggests for the Dutch Wadden Sea a value of 7.0×10^{-5} m^{-1}, to be used in combination with the mean tidal prism,
P = representative tidal prism [m^3].

Qualitatively speaking, a positive correlation between A_c and P makes sense because channels are likely to be larger if they have to convey more water. An obvious disadvantage of the above form is that the constant of proportionality is dimensional.

Measured data from many inlets all over the world have confirmed that such a relationship exists (Fig. 4), although Jarrett (1976) claims that the exponent of P should be 1.1 for the inlets on the US coasts. The validity of Eq. (1) has also been shown for the Dutch and German Wadden Sea inlets (cf. Eysink, 1990; Schroeder *et al.*, 1994). Moreover, a similar relationship was shown to exist over a very wide range of scales for channels inside the flood basin (Gerritsen, 1990; Eysink *et al.*, 1992). In that case, P stands for the total amount of water conveyed by the channel during one tide.

An alternative version of this relationship expresses A_c in terms of the so-called "stability shear stress", which is defined as the shear stress needed to maintain a zero net transport gradient along the channel (Gerritsen, 1990;

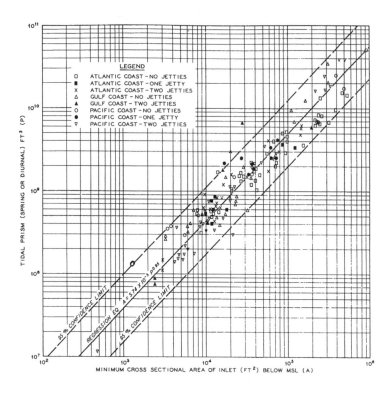

Fig. 4. Empirical relationship between cross-sectional area and tidal prism. (Jarrett, 1976)

Friedrichs, 1995). This shear stress is not necessarily equal to the critical shear stress for sediment motion.

In terms of systems, Eq. (1) represents a typical "black-box" input/output relationship. Unlike a geostatistical model, it describes the state of the system (represented by A_c) as a function of the forcing (represented by P). Clearly, a widely valid empirical relationship like this must concern the equilibrium state, otherwise site-specific history-effects would have led to a much larger scatter in the A_c–P diagram.

Later work on inlet data has revealed a number of other relationships between equilibrium-state properties and the tidal prism, such as:

- the flood-basin volume below MSL (Eysink, 1990), found to be proportional to $P^{1.5}$. This volume is largely equal to the sum of volumes of the flow-conveying channels. Since the cross-sectional area of these channels is

proportional to P and their length probably scales with the square root of the basin area, the proportionality with $P\sqrt{P}$ seems to make sense.
- the sand volume of the outer delta (Walton and Adams, 1976; Eysink et al., 1992), which is proportional to $P^{1.23}$. As the outer delta is mainly formed by the ebb-jet out of the inlet, a relationship with P is to be expected. The weak dependency on the wave conditions, however, is rather surprising. Apparently, the system's sand budget is almost closed and the waves only deform the body of sand;
- the protrusion of the outer delta (De Vriend et al., 1994), which turns out to be proportional to $P^{0.6}$, at least for the Dutch and the East-Frisian Wadden Sea. Note that the effect of the wave conditions cannot be identified here because the wave climate is largely the same throughout the area.

This may give the impression that any large-scale equilibrium-state property of tidal inlets can be expressed in terms of the tidal prism. Clearly, this is only true for the properties which are directly related to the amount of water which moves through the system during the tide.

A very important aspect of the morphological system, viz. the level of the intertidal flats, is mainly correlated to the mean high-water (MHW) level and the local wind/wave climate. In broad outline, shoals tend to be built up by the tide during calm weather and their top is eroded by waves during rough weather (cf. De Vriend et al., 1989).

Another important property which has a rather weak correlation with the prism is the flats' area. By definition, this quantity is equal to the difference between the basin area and the channel area, and the latter is equal to the channel volume divided by some characteristic channel depth. This leads to a relationship of the type

$$A_f = A_b - A_{ch} \approx A_b - \alpha \frac{P\sqrt{A_b}}{D_{ch}} \approx A_b - \beta \frac{H_m}{D_{ch}} A_b^{3/2}, \qquad (2)$$

in which

A_f = the flats' area, i.e., the area above MSL [m^2],
A_b = the gross basin area (flats and channels) [m^2],
A_{ch} = the horizontal area covered by all channels [m^2],
α = constant of proportionality [m^{-1}],
D_{ch} = characteristic channel depth [m],
β = constant of proportionality [m^{-1}],
H_m = mean tidal range [m].

Renger and Partenscky (1974) suggest, on the basis of data from the inlets in the German Bight,

$$A_f = A_b - 0.025\, A_b^{3/2}, \qquad (3)$$

which would mean that the flats' area depends exclusively on the basin area. According to Eq. (2), this can only be the case if D_{ch} is proportional to H_m. Eysink (1990) shows that if a channel-depth relationship exists, it contains a power of P (or H_m) which is significantly less than 1. Apparently, Eq. (3) is an oversimplification of reality.

These properties of the intertidal zone are not only important for the functions of the inlet system (ecosystem, mussel banks), but also for the large-scale morphodynamics. The bathymetric variation in a tidal basin gives rise to a tidal rectification which can contribute significantly to the residual transport (Van de Kreeke and Robaczewska, 1993; Bakker and De Vriend, 1995), and hence, to the net import or export of sediment. Via the so-called "hypsometry effect" (Pethick, 1980; Boon and Byrne, 1981; also see Van Dongeren and De Vriend, 1994), the flats also influence the tidal asymmetry in the gorge: the smaller the flats' area relative to the basin area, the more positive the asymmetry. Thus, the flats regulate the import or export of sediment and enable the system to reach an equilibrium state.

5.2. *Example of application*

The empirical relationships for the large-scale equilibrium state can be utilized to give off-hand predictions of the impact of factors which affect the tidal prism, the basin area, or the MHW level. An example of such a case is the "Friesche Zeegat", one of the inlets of the Dutch Wadden Sea (Fig. 5) with the following characteristics until 1969:

$$H_{m0} = 2 \text{ m}; \; A_{b0} = 2.1\; 10^8 \text{ m}^2; \; P_0 = 3.0\; 10^8 \text{ m}^3;$$
$$A_{c0} = 1.9\; 10^4 \text{ m}^2; \; V_{b0} = 3.5\; 10^8 \text{ m}^3; \; V_{\Delta 0} \approx 1.8\; 10^8 \text{ m}^3; \; A_{f0} = 1.6\; 10^8 \text{ m}^2.$$

In 1969, the Lauwerszee, with an area of $90\; 10^6$ m^2, was separated from the basin by a dam and reclaimed. This means that in 1969 the above values changed abruptly to

$$H_m = 2 \text{ m}; \; A_b = 1.2\; 10^8 \text{ m}^2; \; P = 2.0\; 10^8 \text{ m}^3 \approx 0.65\, P_0;$$
$$A_c = 1.9\; 10^4 \text{ m}^2; \; V_b = 2.7\; 10^8 \text{ m}^3; \; V_\Delta \approx 1.8\; 10^8 \text{ m}^3; \; A_f \approx 0.75\; 10^8 \text{ m}^2.$$

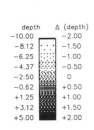
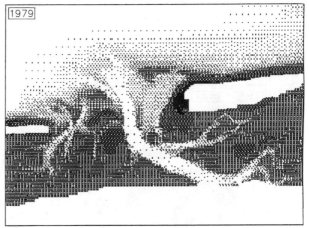

Fig. 5. The Frisian Inlet after closure of the Lauwerszee.

Obviously, the system has responded to this major interference. According to the above relationships, it should seek a new equilibrium state in which

$$H_m = 2 \text{ m}; \quad A_b = 1.2 \; 10^8 \text{ m}^2; \quad P = 2.0 \; 10^8 \text{ m}^3;$$

$$A_c \approx 0.65 \; A_{c0} = 1.2 \; 10^4 \text{ m}^2; \quad V_b \approx (0.65)^{1.5} \; V_{b0} = 1.9 \; 10^8 \text{ m}^3;$$

$$V_\Delta = (0.65)^{1.23} \; V_{\Delta 0} \approx 1.1 \; 10^8 \text{ m}^3; \quad A_f \approx 0.85 \; 10^8 \text{ m}^2 \, .$$

Note that the relative flats' area is expected to increase after closure because the value of this parameter in the remaining basin is initially too small (the Lauwerszee used to be the back part of the basin, and hence, contained a relatively large portion of flats).

In reality, a rapid narrowing of the inlet gorge has taken place, by the formation of a large sandy hook at the east side of the inlet (Fig. 6). The actual amount of sediment import into the basin was initially 7 10^6 m^3/yr, and between 1970 and 1987 it is estimated at 34 10^6 m^3 (Oost and De Haas, 1992). Clearly, the adjustment process has not yet come to an end. Although the delta evolution is rather dependent on long-term meteorological variations, its volume has decreased by 26 10^6 m^3 and its seaward protrusion has decreased roughly by 1.5 km in these 17 years (De Vriend et al., 1994). So the volume decrease is of the same order of magnitude as the sediment demand in the basin. Apparently, the delta serves a buffer for the basin. Yet, the equilibrium values of V_b and V_Δ are proportional to $P^{1.5}$ and $P^{1.23}$, respectively, which means that the basin ultimately demands more sediment than the delta can supply. This sediment has to be provided by the sea bed or the adjacent coasts.

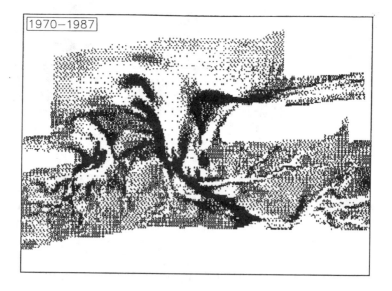

Fig. 6. Frisian Inlet: accretion and ersosion between 1970 and 1987. (The darker the shade, the stronger the accretion.)

5.3. Transient empirical models

Transient empiricial models (O'Connor et al., 1990; Eysink, 1990; Eysink et al., 1992) are one step further along this line. Given the initial state and the equilibrium state, they describe the evolution of the discrepancy between the actual state and the equilibrium state as an exponential decay process. In mathematical terms, this is described by

$$\frac{dA}{dt} = \frac{A_e - A}{\tau}, \qquad (4)$$

in which

A = morphological state parameter (e.g., A_c, V_b, or A_f),
t = time,
A_e = equilibrium value of A,
τ = characteristic time scale of the decay process.

If A_e is a constant, the corresponding solution reads

$$A(t) = A_e - [A_e - A(0)] \exp\left(-\frac{t}{\tau}\right). \qquad (5)$$

If A_e exhibits a slow linear variation in time, for instance in the case of the flats' level responding to mean sea-level rise, the solution reads

$$A(t) = A_e(t) - \tau \frac{dA_e}{dt} - \left[A_e(0) - \tau \frac{dA_e}{dt} - A(0)\right] \exp\left(-\frac{t}{\tau}\right). \qquad (6)$$

The example in Fig. 7 shows that, after the effect of the initial condition has damped out ($t \gg \tau$), the actual state keeps on lagging behind the equilibrium state. This should be borne in mind when trying to find empirical flat-level relationships in areas with a significant rate of sea-level rise.

Transient models of this type primarily concern macroscale phenomena. They rest upon the implicit assumption that each system element behaves independent of the others. In other words, the element would exhibit this behavior if sufficient sediment would be available. In a coherent system like a tidal inlet, this may be true for some elements, but presumably not for all of them. If, for instance, Eq. (4) is utilized to assess the sediment demand of a basin under the influence of a sea-level rise, the interaction with the outer delta and the adjacent island coasts is ignored.

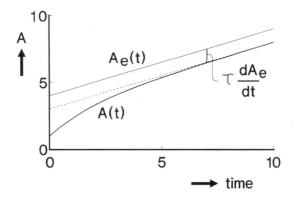

Fig. 7. Flats' level evolution for rising mean high water level.

6. Semi-Empirical, Long-Term Models

6.1. *General approach*

Although a theoretical approach based on formal integration of the constituting equations (see the next chapter) is probably the most accurate, it is a long shot. The identification of the relevant physical processes and the understanding of their nonlinear interaction requires much further research, which is going to take time. Therefore, it seems wise to pursue a more pragmatic line of development at the same time: semi-empirical, long-term modeling. This approach tries to make use of all information available, from the field via the empirical equilibrium-state relationships, and from theory via the large-scale balance equations. Wherever needed, parameterized results of more detailed models are included.

Since the empirical information is not available at a very detailed scale, it is important to think at a high level of aggregation, for instance, in terms of the principal system elements (basin, gorge, outer delta, island coasts). In the following, a brief outline will be given for a number of models for various of these elements.

6.2. *Stability of the gorge*

Wide-open inlets, like those in the Wadden Sea, are not likely to have a strong reducing effect on the tide inside the basin. The tidal amplitude there is closely similar to that at open sea, apart from possible shallow-water effects. If the inlet is narrower, however, it will hamper the tide from entering the basin, with a reduction of the amplitude and the tidal prism as a consequence. But if the tidal prism is reduced, the cross-sectional area of the inlet will also be reduced, which leads to a further reduction of the prism, etc. This phenomenon of positive feedback, which ends up in the closure of the inlet, is usually called *inlet instability*.

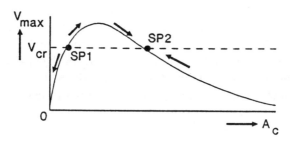

Fig. 8. Stability diagram for tidal inlets.

Escoffier (1940) was the first to point out and model this instability phenomenon. He compared the maximum tidal velocity through the inlet, V_{max}, with the critical velocity for sediment transport in the inlet channel, V_{cr}. V_{max} is a function of the cross-sectional area of the channel, A_c. If A_c is small, this function will be increased because the growth of the tidal prism will be predominant. If A_c is large, the function will be decreased because the the tidal prism has become independent of A_c and the same amount of flow through a wider channel leads to smaller velocities. Figure 8 outlines the curves of V_{max} and V_{cr} against A_c. If $V_{max} = V_{cr}$, which is the case in Fig. 8 for two values of A_c, the system is in equilibrium. The stability of the system in these "stationary points" can be estimated via

$$\Delta(V_{max} - V_c) = \frac{dV_{max}}{dA_c} \Delta(A_c - A_{cs}). \qquad (7)$$

In the stationary point $SP1$, the derivative of V_{max} with respect to A_c is positive, so $V_{max} - V_{cr}$ increases if $A_c - A_{cs}$ increases. On the other hand, A_c tends to increase with V_{max}, which means that, around this stationary point, $V_{max} - V_{cr}$ and $A_c - A_{cs}$ are positively coupled and the system is unstable. Conversely, the system is stable around the other stationary point, $SP2$. All inlets of the Wadden Sea are within this stable domain.

Van de Kreeke (1990a, 1990b) uses a similar technique to analyze the stability of a multiple-inlet system and arrives at the conclusion that a two-inlet system cannot be stable. He suggests that the same conclusion holds for a multiple-inlet system.

6.3. *Di Silvio's basin models*

Di Silvio (1989) proposes a simple box model for the basin, consisting of one deep part which represents all channels and one shallow part which represents all flats (Fig. 9). The total cross-sectional area of the channels and the level of the flats are the dependent variables. The model works with a long-term representative sediment concentration in each part of the model (sea, channel, flat). This concentration is assumed to depend only on local parameters, such as the water depth, but not on the tide. The concentration at sea is assumed to be independent of the state of the basin, so it is a given constant.

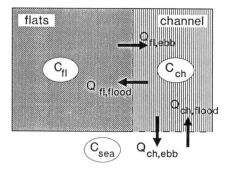

Fig. 9. Principle of Di Silvio's (1989) box model.

The sediment fluxes between the model parts are assumed to be purely advective, which means that they can be written as

$$S_1 = Q_{ch,ebb}\, C_{ch}; \quad S_2 = Q_{ch,flood}\, C_{sea}$$
$$S_3 = Q_{fl,ebb}\, C_{fl}; \quad S_4 = Q_{fl,flood}\, C_{ch}\,, \tag{8}$$

in which

C_{ch}, C_{fl} = representative sediment concentration in the channel and the flats part, respectively,
C_{sea} = representative concentration in the sea part of the model,
$Q_{ch,ebb}$ = total water mass which flows from the channel part to the sea during ebb,

$Q_{fl,ebb}$ = total water mass flowing from the flats part to the channel part during ebb,

etc.

Clearly, the conservation of water mass requires

$$Q_{ch,ebb} = Q_{ch,flood}; \quad Q_{fl,ebb} = Q_{fl,flood}. \tag{9}$$

The sediment balance equations for the channel and the flats part of the model read

$$B_{ch}\frac{dA_{ch}}{dt} = S_1 - S_2 - S_3 + S_4$$
$$= Q_{ch,ebb} C_{ch} - Q_{ch,flood} C_{sea} - Q_{fl,ebb} C_{fl} + Q_{fl,flood} C_{ch}, \tag{10}$$

$$B_{fl}\frac{dz_{fl}}{dt} = S_4 - S_3 = Q_{fl,flood} C_{ch} - Q_{fl,ebb} C_{fl}, \tag{11}$$

in which $A_{ch}(t)$ denotes the total cross-sectional area of the channels and $z_{fl}(t)$ the flats' level. The constants B contain geometrical properties (channel length, flats' area) and the sediment content per unit volume of bed. Combining with Eq. (9), the system is in equilibrium if $C_{fl} = C_{ch} = C_{sea}$.

The sediment dynamics of the system are reflected by a relationship between the concentration and the dependent variable in the relevant model part. In a linear approximation in A_{ch} and z_{fl}, Eqs. (10) and (11) can be elaborated to morphological evolution equations of the form

$$\frac{dA_{ch}}{dt} + X_{ch}A_{ch} = Y_{ch}z_{fl} + Z_{ch}, \tag{12}$$

$$\frac{dz_{fl}}{dt} + X_{fl}z_{fl} = Y_{fl}A_{ch}, \tag{13}$$

in which X, Y, and Z are constants. These coupled linear decay equations form a second-order system, as can be shown by reducing them to

$$\frac{d^2A_{ch}}{dt^2} + (X_{ch} + X_{fl})\frac{dA_{ch}}{dt} + (X_{fl}X_{ch} - Y_{fl}Y_{ch})A_{ch} = 0. \tag{14}$$

If X_{ch}, X_{fl}, and $Y_{fl}Y_{ch}$ are positive, the solution of this equation is of the type

$$A_{ch}(t) = A_{ch,e} + P_1 \exp(-\lambda_1 t) + P_2 \exp(-\lambda_2 t), \tag{15}$$

in which $A_{ch,e}$ is the equilibrium value of A_{ch}, P_1 and P_2 are constants which follow from the initial conditions, and λ_1 and λ_2 are the (real and positive) roots of the characteristic equation of Eq. (14).

Apparently, the dynamic interaction between the flats and the channel parts give rise to a behavior, which differs from the exponential decay, which would be found if the model parts were mutually independent (cf. Sec. 5).

The above model was used with reasonable success to hindcast the morphological evolution of the Venice Lagoon over the last 180 years (Di Silvio, 1991; see Fig. 10). The same author describes a 2D finite-difference version of the concept, in which each grid cell represents a channel or a flats element. The sediment transport in this model is described with an advection/diffusion model, still in terms of the long-term representative concentration C.

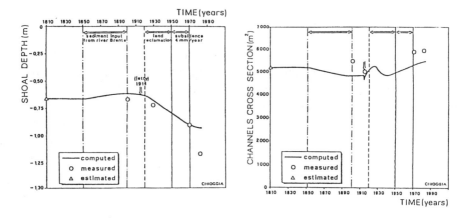

Fig. 10. Hindcast of the morphological evolution of the Chioggia basin (Venice Lagoon) between 1810 and 1990 with Di Silvio's (1989) box model: (a) flats' level, (b) cross-sectional area of the channels. (Di Silvio, 1991)

In either of these models, the constituting equations are reduced to a set of tide-integrated morphological evolution equations which can be solved at the morphological timescale. Thus, the same line of thought is followed as in the formally averaged models which are described in the companion chapter.

6.4. Van Dongeren's basin model

Van Dongeren and De Vriend (1994) suggest the division of a channel/shoal system into a chain of consecutive boxes, aligned according to their distance from the gorge. In each box there is a channel part and a flats part (Fig. 11). The dependent variables in each box are the channel cross-sectional area, A_{ch}, the flats' area, A_{fl}, and the flats' level, z_{fl}. Empirical relationships are used to determine the equilibrium values of these variables.

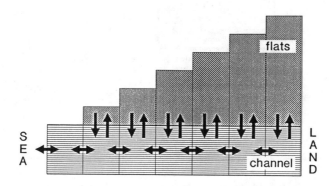

Fig. 11. Principle of Van Dongeren's basin model.

The sediment balance for the channel part of each box involves the exchange with the neighboring channel sections and the adjacent flats. The sediment balance of the flats part involves only the exchange with the adjacent channel part (there is no sediment exchange between flats parts).

In contrast to the models in Subsec. 6.3, the constituting equations are not reduced to one or more morphological evolution equations. In fact, this element of the model is replaced by introducing empirical decay equations for A_{ch} and A_{fl}, which apply if there is enough sediment supply to (or export from) the relevant part of the box:

$$\frac{dA_{ch}}{dt} = \frac{A_{ch,e} - A_{ch}}{\tau_{ch}} \quad \text{and} \quad \frac{dA_{fl}}{dt} = \frac{A_{fl,e} - A_{fl}}{\tau_{fl}} \qquad (16)$$

where $A_{ch,e}$ and $A_{fl,e}$ denote equilibrium values and τ_{ch} and τ_{fl} are characteristic timescales.

The former two quantities follow from empirical relationships, the latter two are input parameters, to be estimated from measured data or otherwise. If there is not enough sediment supply or export, the available supply/export is used for morphological changes, with a fixed sequence of priority (see Fig. 12).

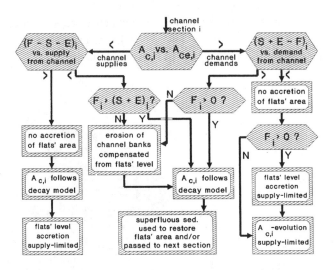

Fig. 12. Flow chart of Van Dongeren's basin model.

The boxes are handled consecutively, in a sequence which corresponds with the direction of propagation of the morphological information ("sand wave"). Sometimes this sequence is rather obvious, like in the case of the Friesche Zeegat (see Sec. 5), but in other cases this is not as clear (cf. Goldenbogen et al., 1994).

The system is essentially driven by the hypsometry effect (see Sec. 5): if the flats' area is too small, the tidal asymmetry in the gorge is positive and sediment is imported. Conversely, sediment is exported if the flats' area is too large. The dynamic response of the basin actually boils down to the tendency of the channel cross-section to adjust rapidly to the hydrodynamic conditions, at the expense of the flats if necessary. In the long run, when the sand (or sand-deficit) wave from the gorge has reached the relevant section, part of the incoming sediment is used to restore the flats.

The spatial resolution along the channel provides the possibility to model bifurcating systems, and to reproduce some of the wave-type behavior of the morphological changes. Yet, the concept is essentially large-scale, and the model results should be treated that way.

This model has been applied with reasonable success to the Friesche Zeegat (see Fig. 13), but in other cases, it yields somewhat poorer results, especially

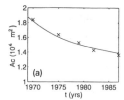 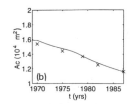 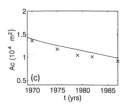

Fig. 13. Hindcast of channel evolution in the Frisian Inlet with Van Dongeren's (1994) basin model, at (a) 3 km, (b) 7 km, (c) 9 km from the gorge. (De Vriend et al., 1994)

for the evolution of the flats' level (Goldenbogen et al., 1994). Clearly, the interaction of flats and channels needs further investigation. Since the concept is limited to basins which are small compared to the tidal wave length (the water-surface elevation has to be in phase throughout the basin), the model in its present form does not work in larger basins. Bakker and De Vriend (1995) describe an alternative network model for such basins.

6.5. Karssen's basin model

Karssen (1994a, 1994b) takes a somewhat similar approach as Di Silvio in a 1D network model concept for estuaries and tidal basins. The basis is a 1D network model for unsteady flow, in which the cross-section of each channel consists of a deep channel part, a low-lying flats part, and a high-lying flats part. The results of a tidal computation with this network model are used to determine a number of equilibrium-state parameters via empirical relationships.

The sediment is assumed to be transported primarily as a suspended load, which means that a representative sediment concentration in each of the three parts of the channel cross-section can be used as a dependent variable. The transport mechanisms in the channel are advection and longitudinal diffusion. The transport on the flats sections is diffusive in the cross-channel direction, i.e., there is no longitudinal transport over the flats. Upon integration of these diffusive fluxes in the cross-channel direction, the net lateral sediment flux between the various parts of a cross-section becomes proportional to the difference in concentration:

$$F_{lc} = D_l h_l \frac{C_l - C_c}{L_{lc}} \quad \text{and} \quad F_{hl} = D_h \frac{h_h}{4} \frac{C_h - C_l}{L_{hl}} \qquad (17)$$

in which

F_{lc} = net sediment flux from the low-lying flats to the channel,
F_{hl} = net sediment flux from the high-lying to low-lying flats,
D_l, D_h = lateral diffusion coefficients for the low-lying and high-lying flats, respectively,
h_l, h_h = effective water depth above these flats,
C_l, C_h = representative sediment concentration above these flats,
C_c = representative sediment concentration in the channel,
L_{lc} = the distance over which the cross-channel integration takes place, so a typical distance from the low-lying flats to the channel, and
L_{hl} = the same from the high-lying to the low-lying flats.

Thus, the concentration evolution equations for the high-lying and the low-lying flats become

$$\frac{\partial (A_h C_h)}{\partial \tau} = W_h w_s (C_{he} - C_h) - F_{hl} \qquad (18)$$

$$\frac{\partial (A_l C_l)}{\partial \tau} = W_l w_s (C_{le} - C_l) - F_{lc} + F_{hl} \qquad (19)$$

in which

A_l, A_h = cross-sectional area on the high-lying and the low-lying flats part, respectively,
W_l, W_h = settling coefficients on these flats,
C_{he}, C_{le} = equilibrium concentration on these flats,
τ = "slow" time coordinate which varies at the morphological time scale,
w_s = sediment settling velocity.

The corresponding sediment balance equation describes the evolution of the cross-sectional areas of the three-channel parts in terms of the net sediment flux from the bed:

$$\frac{\partial A_i}{\partial \tau} = W_i w_s (C_{ie} - C_i) \qquad (20)$$

in which the suffix i ($i = c, l, h$) refers to each of the parts.

Unlike Di Silvio's (1989) model, these sediment balance equations are not elaborated to morphological evolution equations by expressing C_i in terms of A_j ($j = c, l, h$). Instead, Eq. (20) is solved in a simulation loop, together with the concentration equations and the empirical equilibrium-state parameters computed from the tidal model results. After the topography has sufficiently been changed to influence the tide, another hydrodynamic model run is made. To that end, the cross-sectional areas which result from the morphological computation are translated into a seven-point cross-sectional shape per channel section.

A critical point of this model is the definition of the nodal point relationships for the transport (concentration). Depending on how these are chosen, a bifurcated channel system can or cannot persist (Wang et al., 1993). If the transport is distributed according to the flow rates through the channel sections connected to the node for instance, part of the channels will fill up.

A first application of this model to the Friesche Zeegat case (Sec. 5) was rather succesful in forecasting the evolution of the channels' cross-sectional area throughout the basin (Fig. 14). Whether the model performs equally well for the tidal flats remains to be investigated.

6.6. De Vriend et al.'s delta model

The morphological evolution of the basin has its effects on the outer delta. In the Friesche Zeegat case (Sec. 5) the delta volume decreased by about the same amount as was deposited into the basin. Although there is no evidence that it was the sediment from the delta which was deposited into the basin, the amount of sediment involved is so large (approximately $30 \ 10^6 \ m^3$) that this must have been the case to a large extent.

The behavior of the outer delta is determined by tide and sea waves. The tide generally tends to build out the delta seawards, the waves tend to squeeze it shorewards (bulldozer effect). The evolution of the outer deltas in front of the Delta works, in the southwest of The Netherlands, is a spectacular illustration of what happens when one of these agents is eliminated. After closure of the inlets, the deltas were pushed up against the shore and formed large systems of shoals, which seem to be the onset of barrier islands (cf. Stive, 1986; De Vriend and Ribberink, 1988; Steijn et al., 1989). Also the closure of the Lauwerszee led to a spectacular morphological response by the delta of the Friesche Zeegat, viz. the formation of a major sandy hook at the tip of the isle of Schiermonnikoog (see Fig. 6).

These observations illustrate that there are two very powerful, but largely counteracting, agents at work on the outer delta. Any morphodynamic model of the outer delta should therefore take each of these agents into account. A large-scale concept which seems to contain them both is a two-line model in which there is a forced transfer of sediment from the beach line to the inshore

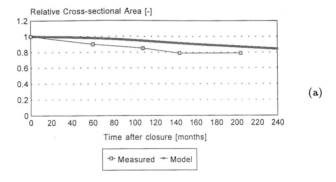

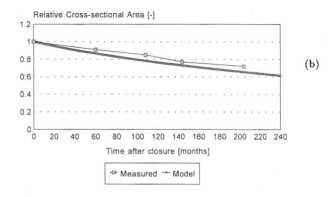

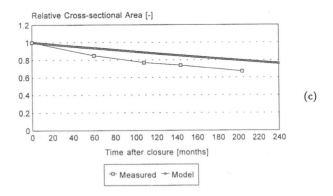

Fig. 14. Hindcast of channel evolution in the Frisian Inlet with Karssen's (1994) basin model at (a) 2.5 km, (b) 7 km, (c) 8 km from the gorge. (Karssen, 1994; reproduced by courtesy of Delft Hydraulics)

line (Fig. 15). This forced cross-shore transport represents the ebb-dominated tidal transport on the outer delta. The "normal" longshore and cross-shore transport components represent the wave-induced effects. The beach and inshore lines respond by taking such an orientation that the wave-induced transport is just enough to feed the forced cross-shore transport at the shoreward end, and to remove the sediment which it delivers at the seaward end. If the forced cross-shore transport is cut off, the system will return to a straight coast.

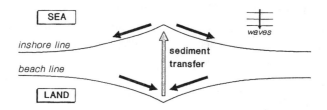

Fig. 15. Delta formation in a two-line model.

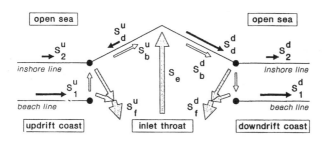

Fig. 16. Principle of De Vriend et al.'s (1994) outer delta model.

De Vriend et al. (1994) describe a model of the outer delta and the adjacent coasts which is based on a refined version of this concept (Fig. 16). It includes not only the main ebb-channel, but also the flood channels near the island tips. The equilibrium state of the delta (volume, seaward protrusion) is described using empirical relationships. The sediment balance is maintained throughout the evolution, and the sediment demand or supply by the basin is taken into account as an input parameter. The model output consists of the cross-shore positions of the beach line and the inshore line as a function of time, and the longshore coordinate, but these results should be interpreted at a higher aggregation level (for instance, in terms of the total sand volume changes in a certain coastal section).

The model was validated against measured data from the Friesche Zeegat (cf. Sec. 5), of which the delta evolution after closure could be reproduced reasonably well (Fig. 17). Further analyses of the results made clear that the delta acts as a mid-term sediment buffer for the basin, but that in the long run the sediment surplus or deficit of the delta has to be compensated by the adjacent island coasts. This indicates that sand mining from the delta can have detrimental long-term effects on the island coasts.

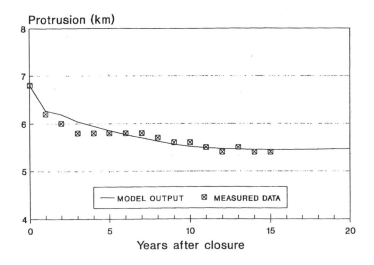

Fig. 17. Hindcast of the outer delta protrusion of the Frisian Inlet with De Vriend et al.'s (1994) model.

6.7. *Steetzel's model of the entire Wadden Sea coast*

Steetzel (1995) uses a similar model to hindcast and predict the large-scale evolution of the entire 150 km-long, multi-inlet coast of the Dutch Wadden

Sea, under the influence of accelerated sea-level rise, climatic change, and subsidence due to oil and gas mining. The response of the basin volume to the relative sea-level rise is assumed to be described by an exponential decay model as given by Eq. (4). The resulting time-evolution of the sediment demand from the basin is determined for each of the inlets. With these time-functions as input conditions at the basin end, the model computes the positions of the island coastlines and the edge of the outer delta.

The basic concept of Steetzel's model is the same as described by De Vriend et al. (1994), but it is based on five lines instead of two in order to be able to deal with the variety of transport mechanisms in the various depth zones. The layers concern the dune front, the subaerial beach, the surf zone, the middle shoreface, and the lower shoreface. They cover the zone between O.D. +7 meters and O.D. −20 meters.

Information from detailed process-based models of the Eierlandse Gat and the Friesche Zeegat is used to set some of the model parameters. The model is validated against 25 years of data from the JARKUS data set (yearly profiles of 800 meters in length, taken at alongshore distances of 250 meters).

Table 1 summarizes the overall sediment balance for the model area, which extends from Callantsoog at the southwest end to the eastern-most tip of Schiermonnikoog at the northeast end. Clearly, the demand from the basins is of predominant importance to the evolution of this coast.

Table 1. Overall sediment balance for the Dutch Wadden Sea coast. (Steetzel, 1995; reproduced by courtesy of Delft Hydraulics)

Model boundary	Sediment flux into the model domain
Callantsoog (southwest end of the model)	+1.1 10^6 m^3/yr alongshore
Schiermonnikoog (northeast end of the model)	−2.1 10^6 m^3/yr alongshore
O.D. −20 m (seaward model boundary)	+ 0.8 10^6 m^3/yr cross-shore
O.D. +7 m (dune top = landward model boundary)	−0.3 10^6 m^3/yr cross-shore
Inlet gorges (demand from basins)	12.7 10^6 m^3/yr cross-shore
Nourishments (sand mined offshore)	+ 2 to 3 10^6 m^3/yr source

A further analysis of JARKUS and other bathymetric survey data has revealed that 75% of the net sediment loss in the area is at the expense of the outer deltas, and that 50–60% takes place at the middle and lower shoreface. Apparently, the erosion is concentrated at the delta edges.

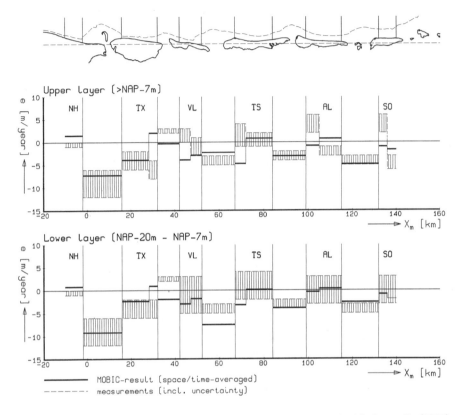

Fig. 18. Long-term evolution of the Wadden Sea coast as predicted with Steetzel's (1995) model.

These findings are confirmed by the model. Figure 18 shows a typical result: the predicted accretion/erosion per coastal section over the next fifty years. The spatial resolution of the presented results is much smaller than that of the numerical model, which computes the cross-shore position of each line as a function of time and longshore position. This aggregation of output is in line with the strong geometrical schematization of the model.

6.8. Compound models

The semi-empirical models which have been discussed so far concern either the basin or the coast (including the outer delta). In both cases, the long-term residual transport through the gorge is prescribed, either as a function of the basin state (basin model) or as a given quantity (delta model). The next logical step would be to couple basin, delta, and island coasts to yield a model for the entire inlet system.

This job is less trivial than it may seem at first sight. Since the transport through the gorge is modeled differently in either case, it is not just a matter of patching. In fact, the processes in the gorge which link the basin to the outer coast should be reanalyzed, taking the influence of all elements of the system into account (for instance, the combined effect of the outer delta and the intertidal flats on the residual transport through the gorge). Process-based simulation models can be of help to such an analysis.

One possible way to couple the basin and the outer coast is to extend the basin model through the gorge and let it interact with the outer coast model. Bakker and De Vriend (1995) suggest the use of a network model for the basin morphology, and to let one or more branches of this network stick through the inlet into the open sea. It must be possible to let the sediment, which is transported through these branches, to be picked up or deposited near the delta edge, thus feeding a multiline coastal model like the one described in the foregoing. If these channels are allowed to migrate, die out, and regenerate, for example, under the influence of the longshore drift, it may even be possible to achieve an essentially higher resolution than under the present models of the outer coast.

At this moment, these are little more than ideas which need to be substantiated, implemented, and tested against reality. Detailed process-based simulation models and frequent (remote-sensing!) topographic surveys can be of great help here.

7. Conclusions

Empirical and semi-empirical models have been shown to be of great value to tidal inlet modeling, although (or maybe because) there is still a long way to go to their understanding in terms of elementary physical processes. The empirical knowledge of inlet systems is based on data from many inlets all over the world. This, together with the vast scientific effort stored in these models, must have removed anomalies from most of the empirical equilibrium-state relationships. Thus, these relationships provide valuable knowledge which should not be ignored, but used to bridge the present gap between process-based simulation and long-term prediction of tidal inlet behavior.

The semi-empirical model approach tries to combine the empirical knowledge with basic physical principles, such as the conservation of sediment and water mass. Thus, it can be a powerful tool to predict the long-term dynamic behavior of tidal inlet systems, as has been shown in various instances. The further development of this approach, together with process-based simulation models, is certainly worth pursuing.

An important research issue in this area is the theoretical substantiation of the empirical equilibrium-state relationships, starting from physical principles. Furthermore, the semi-empirical-model concepts need further improvement and validation. This requires consistent bathymetric records over a sufficiently long period of time (relative to the morphological timescale of interest). Long-lasting monitoring programmes, as well as historical and sedimentological reconstructions, are therefore absolute necessities for further progress. Moreover, better use should be made of process-based simulation models and formally averaged models to improve and refine semi-empirical models.

Acknowledgement

The author wishes to acknowledge the support by Delft Hydraulics, where most of the models described herein were developed and produced. Other important sources of information are the Coastal Genesis project, sponsored by the Netherlands Ministry of Transport and Public Works (Rijkswaterstaat), and the ISOS*2-WADE project, which is sponsored jointly by Rijkswaterstaat and the German Ministry of Science and Technology. Also colleagues at the Forschungsstelle Kueste, Norderney, are gratefully acknowledged for their willingness to provide information and ideas.

References

Bakker, W. T. and H. J. De Vriend (1995). Resonance and morphological stability of tidal basins. *Mar. Geol.* 126: 5–18.

Bakker, W. T. and D. S. Joustra (1970). The history of the Dutch coast in the last century. *Proc. 12th Int. Conf. Coastal Eng.* ASCE. **26**: 709–728.

Boon, J. D. III and R. J. Byrne (1981). On basin hypsometry and the morphodynamic response of coastal inlet systems. *Mar. Geol.* **40**: 27–48.

De Vriend, H. J. (1991). Mathematical modeling of large-scale coastal behavior. *J. Hydraul. Res.* **29** (6): 727–753.

De Vriend, H. J. and J. S. Ribberink (1988). A quasi-3D mathematical model of coastal morphology. *Coastal Engineering 1988 Proc.*, ed. B. L. Edge. ASCE. 1689–1703.

De Vriend, H. J., W. T. Bakker and D. P. Bilse (1994). A morphological behavior model for the outer delta of mixed-energy tidal inlets. *Coastal Eng.* **23** (3&4): 305–327.

De Vriend, H. J., M. Capobianco, T. Chesher, H. E. De Swart, B. Latteux and M. J. F. Stive (1993). Approaches to long-term modeling of coastal morphology: A review. *Coastal Eng.* **23** (1-3): 225–269.

De Vriend, H. J., T. Louters, F. Berben and R. C. Steijn (1989). Hybrid prediction of a sandy shoal evolution in a mesotidal estuary. *Hydraulic and Environmental Modeling in Coastal, Estuarine and River Waters*, eds. R. A. Falconer, P. Goodwin and R. G. S. Matthew, Gower Technical, Aldershot, UK. 145–156.

Di Silvio, G. (1989). Modeling the morphological evolution of tidal lagoons and their equilibrium configurations. *Proc. 23rd IAHR Congress*, Ottawa, Canada. C.169–C.175.

Di Silvio, G. (1991). Averaging operations in sediment transport modeling; short-step versus long-step morphological simulations. *The Transport of Suspended Sediment and its Mathematical Modeling, Int. IAHR/USF Symp.*, Florence, Italy. 723–739 (preprints).

Ehlers, J. (1988). *The Morphodynamics of the Wadden Sea.* Balkema. Rotterdam. 397.

Escoffier, F. F. (1940). The stability of tidal inlets. *Shore and Beach* **8** (4): 114–115.

Eysink, W. D. (1990). Morphologic response of tidal basins to changes. *Coastal Engineering 1990 Proc.*, ed. B. L. Edge. ASCE. 1948–1961.

Eysink, W. D., E. J. Biegel and F. J. M. Hoozemans (1992). Impact of sea level rise on the morphology of the Wadden Sea in the scope of its ecological function; investigations on empirical morphological relations, Delft Hydraulic, ISOS*2 Report H 1300. 73.

Fenster, M. S., R. Dolan and J. F. Elder (1993). A new method for predicting shoreline positions from historical data. *J. Coastal Res.* **9** (1): 147–171.

Friedrichs, C. T. (1995). Stability shear stress and equilibrium cross-sectional geometry of sheltered tidal channels. *J. Coastal Res.* **11** (2) (in press).

Gerritsen, F. (1990). Morphological stability of inlets and channels in the Western Wadden Sea. Rijkswaterstaat, Report GWAO-90.019.

Goldenbogen, R., E. Schroeder, H. Kunz and H. D. Niemeyer (1994). Intermediate report on the "WADE" research project. Niedersachsisches Landesamt fuer Oekologie, Forschungsstelle Kueste, Report MTK 0508. 111 (in German).

Hayes, M. O. (1979). Barrier island morphology as a function of tidal and wave regime. *Barrier Islands from the Gulf of St. Lawrence to the Gulf of Mexico*, ed. S. P. Leatherman. Academic Press. 1–27.

Jarrett, J. T. (1976). Tidal prism area relationships. US Army Corps of Engineers, GITI Report No. 3. 32.

Karssen, B. (1994a). A dynamic/empirical model for the long-term morphological development of estuaries; Part II: Development of the model, Phase I. Delft Hydraulics, DYNASTAR Report Z 715-I. 31.

Karssen, B. (1994b). A dynamic/empirical model for the long-term morphological development of estuaries; Part II: Development of the model, Phase II. Delft Hydraulics, DYNASTAR Report Z 715-II. 38.

Medina, R., C. Vidal, M. A. Losada and I. G. Losada (1992). Three-mode principal component analysis of bathymetric data applied to Playa Castilla (Huelva, Spain). *Coastal Engineering 1992*, ed. B. L. Edge. ASCE. 2265–2278.

O'Brien, M. P. (1931). Estuary tidal prism related to entrance areas. *Civ. Eng.* **1** (8): 738–739.

O'Brien, M. P. (1969). Equilibrium flow areas of inlets on sandy coasts. *J. Wtrwy. Harbors Div., ASCE* **95** (WW1): 43–52.

O'Connor, B. A., J. Nicholson and R. Rayner (1990). Estuary geometry as a function of tidal range. *Coastal Engineering 1990 Proc.*, ed. B. L. Edge. ASCE. 3050–3062.

Oost, A. P. (1995). The cyclic development of the Pinkegat Inlet system and the Engelsmanplaat/Smeriggat, Dutch Wadden Sea, over the period 1832-1991. *Dynamics and Sedimentary Developments of the Dutch Wadden Sea, with Special Emphasis on the Frisian Inlet*, Doctoral Thesis, Utrecht University (in press).

Oost, A. P. and H. de Haas (1992). The Frisian Inlet, morphological and sedimentological changes in the period 1970–1987. Utrecht University, Institute of Earth Sciences, Coastal Genesis Report. 68 (in Dutch).

Oost, A. P. and H. de Haas (1993). The Frisian Inlet, morphological and sedimentological changes in the period 1927–1970. Utrecht University, Institute of Earth Sciences, Coastal Genesis Report. 94 (in Dutch).

Oost, A. P., H. de Haas, F. IJnsen, J. M. Van den Boogert and P. L. De Boer (1993). The 18.6 year nodal cycle and its impact on tidal sedimentation. *Sediment Geol.* **87**: 1–11.

Pethick, J. S. (1980). Velocity surges and asymmetry in tidal channels. *Est. Coastal Mar. Sci.* **11**: 321–345.

Preisendorfer, R. W. and C. D. Mobley (1988). *Principal Component Analysis in Meteorology and Oceanography.* Elsevier. 425.

Renger, E. and H. W. Partenscky (1974). Stability criteria for tidal basins. *Proc. 14th Coastal Eng. Conf.*, Copenhagen, Denmark. ASCE. 1605–1618.

Schroeder, E., R. Goldenbogen and H. Kunz (1994). Parametrization for conceptual morphodynamic models of Wadden Sea areas. *Coastal Engineering 1994*, ed. B. L. Edge. ASCE (in press).

Steetzel, H. J. (1995). Modeling of the interrupted coast; prediction of the coastline and the outer deltas of the Wadden Sea coast over the period 1990–2040. Delft Hydraulics, Coastal Genesis Report H 1887. (in Dutch).

Steijn, R. C., T. Louters, A. J. F. Van der Spek and H. J. De Vriend (1989). Numerical model hindcast of the ebb-tidal delta evolution in front of the Deltaworks. *Hydraulic and Environmental Modeling of Coastal Estuarine and River Waters*, eds. R. A. Falconer et al., Gower Technical, Aldershot, UK. 255–264.

Stive, M. J. F. (1986). A model for cross-shore sediment transport. *Coastal Engineering 1986 Proc.*, ed. B. L. Edge. ASCE. 1551–1564.

Van de Kreeke, J. (1990a). Stability analysis of a two-inlet bay system, *Coastal Eng.* **14**: 481–497.

Van de Kreeke, J. (1990b). Can multiple inlets be stable? *Est., Coastal Shelf Sci.* **30**: 261–273.

Van de Kreeke, J. and K. Robaczewska (1993). Tide-induced transport of coarse sediment: Application to the Ems estuary. *Neth. J. Sea Res.* **31** (3): 209–220.

Van Dongeren, A. R. and H. J. De Vriend (1994). A model of morphological behavior of tidal basins. *Coastal Eng.* **22** (3&4): 287–310.

Walton, T. L. and W. D. Adams (1976). Capacity of inlet outer bars to store sand. *Proc. 15th Coastal Eng. Conf.*, Honolulu, USA. ASCE. 1919–1937.

Wang, Z. B., R. J. Fokkink and B. Karssen (1993). Theoretical analysis on nodal point relations in one-dimensional morphodynamic models. Delft Hydraulics, Report Z 473-II. 26.

Wijnberg, K. and F. C. J. Wolf (1994). Three-dimensional behavior of a multiple bar system, *Coastal Dynamics '94*, eds. A. S.-Arcilla, M. J. F. Stive and N. C. Kraus. ASCE. 59–73.

MATHEMATICAL MODELING OF MESO-TIDAL BARRIER ISLAND COASTS
Part II: PROCESS-BASED SIMULATION MODELS

HUIB J. DE VRIEND and JAN S. RIBBERINK

The present chapter concerns the class of so-called process-based simulation models for large meso-tidal inlets. Together with the previous chapter (Part I), it gives an overview of the variety of mathematical model concepts in use for this type of problem.

Theoretical and modeling aspects of Initial Sedimentation/Erosion models (ISE models) and Medium-Term Morphodynamic models (MTM models) are discussed, and some key points of modeling practice are indicated. For each model type, example applications are described. Finally, some important research isssues are identified.

1. Introduction

Mathematical modeling of tidal inlets can be based on either of the following strategies (cf., Terwindt and Battjes, 1990):

- down-scaling, which starts from large-scale field information on the system's behavior and tries to unravel the underlying physical processes only as far as what is needed for good predictions. This approach is followed in most of the empirical and semi-empirical models which are discussed in the previous chapter.
- up-scaling, which starts from the observation that tidal inlet behavior, however largescale, is the effect of water and sediment motion. Hence, it must be possible to simulate this behavior with a model based on descriptions of the elementary hydrodynamic and sediment transport processes. The constituting equations of such a model are derived from conservation laws for wave energy, wave number, water mass, mean flow momentum, and sediment mass. This process-oriented approach is the key element of the models which will be discussed hereafter.

Present-day computer capacity and the development of powerful software systems for mathematical modeling of waves, currents, and sediment transport have brought process-based mathematical simulation models of morphological evolutions within the reach of tidal inlet research. Yet, the literature on applications of this type of models to tidal inlets is far from abundant.

A possible explanation is that tidal inlets are extremely complex dynamic systems with a variety of processes and mechanisms. At this moment, they are insufficiently understood to be sure that the elementary physics incorporated in a simulation model is sufficient to reproduce the phenomena of interest. Any simulation model, at one scale level or another, has to call upon "black-box" empirical or parametric submodels (cf., turbulence modeling, or wave-driven current modeling). This implies simplifications and approximations of which the long-term effects on the morphodynamic behavior can hardly be overseen.

Process-based simulation modeling has not yet reached the stage in which one can claim that "everything" is in the model, and that it takes only a big computer to simulate large-scale behavior.

2. ISE Models Versus Morphodynamic Models

Process-based simulation models consist of a number of modules which describe waves, currents, and sediment transport, respectively. Depending on whether the dynamic interaction of these processes with the bed-topography changes is taken into account, these modules are used sequentially or in a time-loop (see Figs. 1(a) and (b), respectively). Models of this type essentially work at the hydrodynamic timescale (typically one tidal period). Larger timescales are reached by continuing the computations over a longer time-span.

Models which take the interaction with the topographic changes into account are called Medium-Term Morphodynamic models (MTM models). In principle, they are able to describe the mesoscale dynamic behavior of a morphological system, such as the formation and migration of morphological features (channel shoaling or deepening, bar formation and migration, sandwaves, etc.). Models which do not take this interaction into account only describe microscale phenomena, such as the sedimentation/erosion rate on a given bed topography. In the time-evolution of the system's state, this represents the initial changes. Therefore, this type of models is called Initial Sedimentation/Erosion models (ISE models).

The above terminology is relative to the inherent morphodynamic timescale of the phenomena of interest. The term "initial" therefore does not imply that the results of such a model only provide a snapshot of the hydrodynamic timescale. In the case of large-scale evolution of an inlet system, for instance, the morphological timescale can be of the order of decades or centuries. Morphological changes over a few years at this large spatial scale may therefore be considered as "initial". The essence of ISE models is that the feedback of morphological changes on the water and sediment motion is negligible.

This may explain why, in spite of the inherent limitations of ISE models, there is a demand for their practical application, especially to 3D cases. Most of the practical tidal inlet problems belong to this category. The model is utilized to investigate wave, currents, and sediment transport fields in the area, and their response to projected interferences (engineering works, sand mining concessions, land reclamation projects, etc.). An example of such a case will be discussed later.

Both the ISE and the MTM modeling processes consist of a number of steps, such as problem definition, formulation of objectives, analysis of physical processes, model set-up, model composition, input schematization and definition of runs, processing and visualization of results, verification, interpretation and assessment of results, etc. In the following sections, some of these steps will be discussed in depth, with the emphasis on specific aspects of tidal inlet modeling.

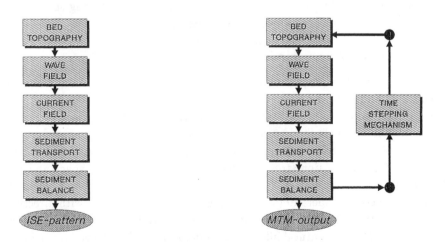

Fig. 1. Aggregated flow charts of process-based simulation models for tidal inlets: (a) Initial Sedimentation/Erosion models, and (b) Medium-Term Morphodynamic models.

3. Model Composition

3.1. *General*

Both ISE and MTM models require a careful selection of the constituent models. Clearly, each of the relevant constituent processes has to be modeled adequately. This adequacy, however, is not only a matter of the process description *per se*, but also of the combination of modules which form the model as a whole. If, for instance, the wave model gives reasonable predictions of the wave height distribution, and the tidal model has been shown to reproduce water levels and current velocities within acceptable bounds, their combination does not necessarily give good results for the near-bed velocity which is put into the sediment transport model.

Especially in complex situations such as tidal inlets, the model composition ought to be the result of an analysis which starts from the objective of the study and its translation into phenomena to be modeled. This analysis should identify the relevant underlying processes and mechanisms, and assess the type of constituent models which is needed to describe them. The combination of constituent models ultimately chosen should be balanced from an efficiency and accuracy point of view, and it should not lead to spurious interactions or a blow-up of inaccuracies (cf., De Vriend, 1987).

Tidal inlets involve a number of specific aspects which need due consideration when modeling the constituent processes. In the following sections, the most important of these aspects will be discussed.

3.2. *Aspects of wave modeling*

The complex topography of the outer delta, with one or more deep ebb-channels and often a well-developed channel/shoal system on the fringe, leads to a much more complex wave field than an uninterrupted piece of exposed coast. The channels tend to convey the wave energy, and the shoals give rise to complex refraction patterns. Wave breaking and the associated current-driving force (radiation-stress gradient) are quite variable in space and drive complex circulation patterns (Fig. 2). In order to describe these phenomena, a spectral or semi-spectral (single-frequency) refraction model with directional spreading (the HISWA model, for instance; see Holthuysen *et al.*, 1989) will usually do for wide natural inlets. The computation of the wave-induced forces via the radiation stresses requires caution in order to avoid large numerical errors (cf., Dingemans *et al.*, 1987).

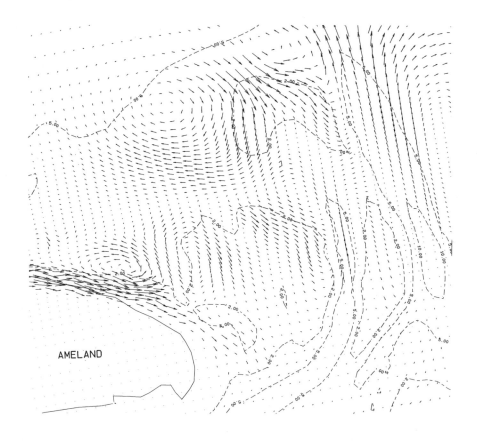

Fig. 2. Residual circulation patterns on the outer delta of the Frisian Inlet. (Steijn and Hartsuiker, 1992; reproduced by courtesy of Delft Hydraulics)

Inlets which are fixed by engineering works, such as jetties and revetments, are usually narrower (inlets in the East Coast of the USA) and more reflective. This can lead to very complex wave patterns which can only be described if reflection and diffraction are included (cf., Kirby et al., 1994).

Wave penetration into the basin mainly occurs via the deeper channels (Fig. 3), where the waves experience less bottom friction and where they tend to be trapped by refraction. For the wide inlets considered herein, these phenomena are described reasonably well with the aforementioned type of (semi-)spectral refraction model.

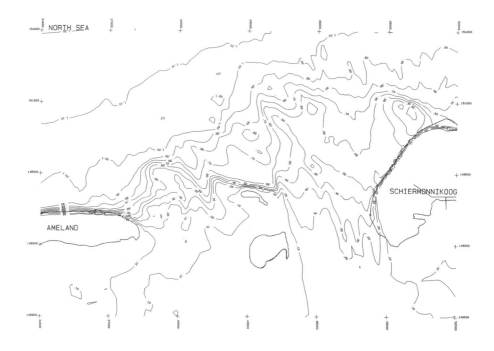

Fig. 3. Wave penetration into the Frisian Inlet. (Steijn and Hartsuiker, 1992; reproduced by courtesy of Delft Hydraulics)

The top level of the tidal flats inside the basin is the result of two opposing agents, viz. tide (accretive) and wave action (erosive) (see De Vriend et al., 1989). Since this level as well as the wave conditions on the flats are important factors to the ecosystem; the ability to adequately predict the waves in the intertidal zone will be a requirement in many applications. This means that the model has to include wave generation. Depending on the situation (exposure to sea waves, wind-fetch inside the basin), it may also be necessary for the wave model to describe sea waves and locally generated waves together, i.e., to allow for a double-peaked spectrum.

3.3. *Aspects of current modeling*

As sediment transport responds nonlinearly to the current velocity, the net transport through an inlet is quite sensitive to asymmetries in the tidal velocity

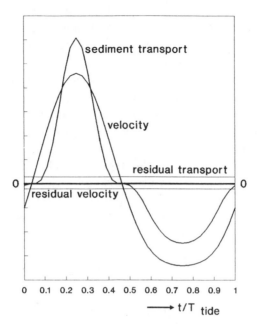

Fig. 4. Tidal asymmetry, residual current, and residual transport.

and stage curves (Fig. 4). One type of asymmetry concerns higher harmonics (overtides), which can be due to shallow-water effects, or to asymmetric flooding and drying of the intertidal zone (hypsometry effect) (cf., Pethick, 1980; Speer and Aubrey, 1985; Friedrichs and Aubrey, 1994). Another type concerns tidal rectification (subharmonics, residual currents), for instance, due to inertial effects (ebb-jet), coriolis effects (ebb- and flood-branches of a channel), and topography-induced rectification (Bakker and De Vriend, 1995).

The latter phenomenon is related to the phase coupling between velocity and water level, due to which the residual current above low water tends to be flood-dominated, and below low water, ebb-dominated. As the transport of sand through the channels is determined mainly by the near-bed velocity, this mechanism tends to export sand through the channels.

Also, storm surges can give rise to important asymmetries of either type. Although current measurements under storm-surge conditions are scarce, observations by fishermen suggest that surge-induced currents through the inlet can be so strong, that the tide does not even turn. Bathymetric surveys before and after such an event show large morphological changes, but if the inlet is sufficiently stable, these effects tend to be annihilated within a few months (Niemeyer, 1994). If the inlet is less stable, a major event can bring it into an essentially different state from which it does not return.

The relative importance of these various sources of asymmetry depends on the situation and the timescale considered. Van de Kreeke and Robaczewska (1993) show for the tide-dominated Ems-Dollard estuary that the contribution of the residual current to the long-term residual transport is even larger than that of the overtides.

A carefully calibrated, 2D, depth-integrated model for combined tidal and wave-driven currents generally suffices to describe these phenomena, provided that it contains a flooding and drying procedure which takes due account of the hydrodynamics involved (i.e., not only based on mass conservation). In formulae, such a model reads

$$\frac{DU}{Dt} - f_c \, Vh = -gh \frac{\partial \zeta}{\partial x} + \frac{\tau_{wx}}{\rho} + \frac{F_x}{\rho} - \frac{\tau_{bx}}{\rho} + HDT \qquad (1)$$

$$\frac{DV}{Dt} + f_c \, U = -gh \frac{\partial \zeta}{\partial y} + \frac{\tau_{wy}}{\rho} + \frac{F_y}{\rho} - \frac{\tau_{by}}{\rho} + HDT \qquad (2)$$

$$\frac{\partial \zeta}{\partial t} + \frac{\partial}{\partial x}\left(Uh + \frac{M_x}{\rho}\right) + \frac{\partial}{\partial y}\left(Vh + \frac{M_y}{\rho}\right) = 0 \qquad (3)$$

in which

- t = time,
- x, y = horizontal cartesian coordinates,
- U, V = x and y component of the depth-averaged velocity,
- h = water depth below MSL,
- ζ = water surface elevation above MSL,
- D/Dt = material derivative, moving along with the depth-averaged flow,
- f_c = coriolis parameter,
- g = acceleration due to gravity,
- ρ = mass density of the fluid,
- τ_{wx}, τ_{wy} = x and y component of the wind shear stress,
- F_x, F_y = x and y component of the wave-induced effective stress,
- τ_{bx}, τ_{by} = x and y component of the bottom shear stress,
- M_x, M_y = x and y component of the wave-induced mass flux,
- HD = horizontal diffusion terms.

Note that the wave effects are included via the wave-induced effective stress, the bottom shear stress, and the wave-induced mass flux. The latter usually has a rather small contribution to the total velocity. It has to be included, however, if the wave-induced undertow is taken into account, e.g., via a $2\frac{1}{2}D$ approach (see below).

Various types of 3D currents are common in tidal inlets. Channel curvature and coriolis effects give rise to secondary circulations in the vertical plane, and wave-induced mass flux and undertow also play a part. These secondary flow components usually have a relatively weak contribution to the instantaneous sediment transport, but some of them are not changing signs during the tide, whence their relative contribution to the residual transport can be significant. Moreover, if these residual transport components have to be compensated by the downslope gravitational transport, their morphological impact can be large. An example of such a case is the formation of a point bar near the inner bank of a curved channel, to compensate for the net centripetal transport component due to the curvature-induced secondary flow.

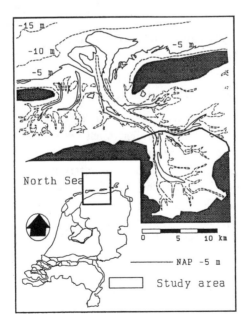

Fig. 5. MTM model of the Frisian Inlet.

Wang et al. (1991) investigate the role of secondary flow effects in a medium-term morphodynamic model (tide only) of the inlet Het Friesche Zeegat (Frisian Inlet) in the Dutch Wadden Sea (see Fig. 5). They find that secondary flows, curvature-induced as well as coriolis-induced, cannot be disregarded in such a situation. The morphological evolution near the tip of the updrift barrier island (Ameland), for instance, is predicted incorrectly if the curvature-induced secondary flow is disregarded. Also, channel migration is likely to be influenced by secondary flow effects.

A rather efficient way to include these effects is a $2\frac{1}{2}$D approach, in which the results of a 2D, depth-averaged flow computation are used in 1D vertical profile models for the main velocity and the secondary flow (Kalkwijk and Booij, 1986). Depending on the input needed for the sediment transport module, this yields the total bottom shear stress or the total near-bed velocity as a function of space and time. This approach only works if the secondary flow velocity is much weaker than the main velocity. A more elegant and robust version is the *quasi-3D* approach, in which the 1D profile models and the 2D, depth-averaged model are intrinsically coupled (see Zitman, 1992, for tidal currents; also see De Vriend and Stive, 1987, for wave-induced currents). Examples of $2\frac{1}{2}$D and quasi-3D model applications to tidal inlets are described by Wang et al. (1991, 1995) for the basin and gorge, and De Vriend and Ribberink (1988) for the outer delta.

3.4. *Aspects of sediment transport modeling*

Sediment transport can occur in various modes, roughly divided into bed load and suspended load. Sediment grains which are transported in the bed-load mode are in more or less continuous contact with the bed (sliding, rolling, jumping, fluidized), and they often move intermittently between periods of being trapped in bed forms. In the suspended-load mode, the grains are held in suspension by turbulent diffusion.

Modelers usually make a slightly different distinction, viz. between transport formulae and advection/diffusion models. Each of the largely empirical transport formulae relates the transport rate (bed load and some forms of suspended load) at a point to local parameters, such as near-bed velocity, bed shear stress, critical shear stress, grain size, etc. Horikawa (1988) and Van Rijn (1989) give reviews of transport formulae for fluvial, estuarine, and coastal applications.

In the case of a horizontal bed, the transport direction is assumed to coincide with the direction of the bed shear stress. If the bed is sloping, a downslope gravitational transport component deflects the transport direction. This component usually forms only a small part of the total transport. Nevertheless it can have strong morphological effects, as stated before. Even if this is not the case, it is worthwhile to take this component into account in morphodynamic models, since it represents a natural smoothing effect on bottom perturbations (cf., De Vriend et al., 1993b).

The current-induced transport rate according to a transport formula is an odd nonlinear function of the near-bed current velocity. Thus, a periodic velocity can give rise to a residual transport if there are higher harmonics. This residual transport has very little to do with the residual current velocity, it may even be in the opposite direction (Fig. 4). If at a point all tidal constituents are given, and the M_2 component is predominant, it can be shown that only M constituents contribute significantly to the long-term residual transport (Van de Kreeke and Robaczewska, 1993). All other constituents have a frequency which differs from an exact multiple of the M_2 frequency, whence their interaction with the M_2 constituent leads to time-modulations and a zero mean contribution in the long run.

Yet, it is not sufficient in a tidal inlet model to impose the net current, the predominant tidal constituent, and its overtides at the open boundaries. Due to the complex flooding and drying topography, each constituent generates residual currents which do contribute to the residual transport.

Advection/diffusion models for suspended load transport usually concern the sediment concentration. Their basic form reads

$$\frac{\partial c}{\partial t} + u \cdot (\nabla c) - w_s \cdot (\nabla c) = \nabla \cdot (D \nabla c) \tag{4}$$

in which

c = sediment concentration, by weight or by volume,
∇ = gradient vector,
u = current velocity vector,
w_s = settling velocity vector,
D = turbulent diffusion coefficient.

These models reflect that in the horizontally uniform equilibrium state (zero gradients in time and horizontal space), the upward diffusive flux is balanced by the downward settling flux. Away from the equilibrium state, an observer who moves along with the current sees a tendency of the concentration to adjust to the equilibrium state. This means that the transport has a nonlocal character in space and time: The concentration at a given point depends on its history further upstream. The characteristic length and timescales involved in these lag effects can be estimated by (cf., Wang, 1992; Katopodi and Ribberink, 1992)

$$L_c = O\left(\frac{|u|h}{w_s}\right) \quad \text{and} \quad T_c = O\left(\frac{h}{w_s}\right) \tag{5}$$

in which

L_c = characteristic decay-length,
T_c = characteristic decay-time,
h = water depth,
$|u|$ = a representative measure of the current velocity,
w_s = magnitude of the settling velocity.

The characteristic timescale for sand with a settling velocity in the order of 10^{-2} m/sec is usually less than 15 minutes, so much smaller than the the tidal period. The corresponding length scale is of the order of a kilometer, which is not small relative to the dimensions of channels and shoals. Spatial lag effects should therefore be taken into account in a suspended-sediment model for such a case.

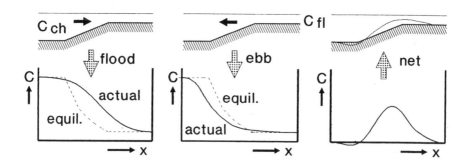

Fig. 6. Shoal formation by tidal transport.

This spatial lag effect may explain how the tide can build up a shoal (Fig. 6). When the flats are flooded, sediment-laden water from the channels moves up the shoals, where the velocity drops and the sediment tends to settle. Due to the lag effect, the sediment is brought further up the flats. When the tide turns and the flats are drying, the water which comes from the flats contains little sediment when it reaches the channels. There the concentration tends to grow, but due to the lag effect less sediment is brought back into the channels during ebb then has been carried up the flats during flood. This picture reverses in the presence of waves, which are much more effective in stirring sediment on the flooded flats than in the deep channels. In that case, the ebb-flow from the flats contains more sediment than the flood flow towards them and the flats are eroded. Since accretion and erosion occur at different times, flats can only be in dynamic equilibrium.

The settling velocity of fine cohesive sediment is much smaller than that of sand. Hence, the characteristic timescale of the adjustment process is of the order of magnitude of the tidal period or larger, and the length scale is of the order of the basin size or larger. An advection/diffusion model is therefore essential to describe the transport of this type of sediment. As a consequence of these large characteristic scales, the residual transport is determined by other properties of the tidal motion than in the case of sand (cf., Dronkers, 1986).

Apart from this, cohesive sediment behaves quite differently from sand (for instance, see Berlamont et al., 1993, and Teisson et al., 1993). One specific complication of cohesive sediment is consolidation, which makes erodibility a function of the time elapsed since deposition. In a tidal environment, this introduces a time-hysteresis which can lead to residual deposition (cf., Fritsch et al., 1989). In order to take this into account, a model should include an administration of the time since deposition of different bottom layers.

Lag effects in suspended sand transport are sometimes taken into acccount via the depth-averaged equation

$$\frac{\partial(h\bar{c})}{\partial t} + \gamma \frac{\partial(\bar{u}h\bar{c})}{\partial x} + \gamma \frac{\partial(\bar{v}h\bar{c})}{\partial y} = -\alpha(\bar{c} - \bar{c}_e) \qquad (6)$$

in which the overbars denote depth-averaging and

x, y = horizontal cartesian coordinates,
u, v = velocity components in x and y direction, respectively,
γ = ratio between the depth-averaged sediment flux and the product of the depth-averaged velocity and the depth-averaged concentration,
α = decay coefficient,
c_e = equilibrium concentration, given as a function of flow and sediment parameters.

This is a very crude approximation which does not do much justice to the physical processes involved. A quasi-3D approach yields a physically much better founded equation of the type (Wang, 1992; Katopodi and Ribberink, 1992)

$$T_c \frac{\partial \bar{c}}{\partial t} + \frac{L_c}{|\bar{u}|} \left[\bar{u} \frac{\partial \bar{c}}{\partial x} + \bar{v} \frac{\partial \bar{c}}{\partial y} \right] = \frac{\alpha}{h}(\bar{c} - \bar{c}_e) \qquad (7)$$

in which the decay parameters T_c, L_c, and α follow from a 1D vertical model of the sediment concentration profile. An alternative to this quasi-3D approach is to solve the complete 3D concentration equation (4). Van Rijn et al. (1991) describe such a model for current-borne suspended sediment.

The above models concern the wave-averaged current-borne transport. They may include the net stirring effects of waves, but they ignore typical wave-related mechanisms such as the residual wave-borne transport due to the asymmetry of the wave orbital motion, and the transport due to the wave-induced boundary layer streaming. These mechanisms play a part in wave-dominated environments like the outer delta and the adjacent island coasts. Too often, they are lumped under the heading "cross-shore transport mechanisms" and ignored as being too difficult to model.

In recent years, coastal morphological modelers have started to open this "black box", trying to simulate coastal profile evolution (cf., Roelvink and Broker, 1993). A fairly crude way to take these effects into account is to apply an intra-wave transport formula (e.g., Bailard, 1981). A more sophisticated approach is to model the suspended sediment concentration at the intra-wave level (Deigaard et al., 1986; also see Fredsøe and Deigaard, 1992). Obviously, it is not efficient to simulate tidal inlet morphodynamics at an intra-wave timescale. This problem can be overcome by running the intra-wave transport model for the full range of possible inputs and tabulating the resulting residual transport rates and directions as functions of input parameters, such as wave height, wave direction, wave period, current velocity, etc. This table replaces the transport module in the inlet simulation model (cf., Brøker Hedegaard et al., 1993).

4. Input Schematization and Definition of Model Runs

The natural variability of input conditions, such as waves, wind stress, and storm surges, poses a special problem to the practical operation of process-based simulation models. Firstly, running the model with real-time input is seldomly feasible, or even impossible if it concerns the future. In the case of ISE models, running input time-series other than the tide does not make much sense. In the case of MTM models, which can have long intervals between two consecutive hydrodynamic updates with the complete wave and current models, the "pace" of the computation process seldom coincides with the frequency of variation of the input conditions.

Therefore, the randomly varying input conditions have to be replaced by representative steady conditions (waves, wind), or a representative storm surge. This input schematization leads to a manageable number of input sets, each with its own probability of occurrence. The model is run for each set, and the results are weighted according to the probability of occurrence of the input p_i. In the case of sediment transport, the nonlinearity of the system leads to weight factors which differ from p_i (cf., De Vriend, 1994). Criteria to determine these factors are, for example, the correct representation of the longshore drift on a given part of the coast, or the correct representation of the tidal transport capacity in another point of the model domain (cf., De Vriend et al., 1993a).

Note that the method of input schematization, the model runs to be made, and the weighted composition of the results of the runs to predictions are mutually coupled. If the schematization principle changes, the weights will also change.

5. Validation

Validation is a continuous process during the entire course of a modeling project. It involves not only the verification of model results against measured data, but also other activities which contribute to the user's control of the model results, such as analyses of the system's behavior in reality and according to the model, analyses of the effects of model simplifications, execution of test-runs on simple and verifiable cases, sensitivity analyses, etc.

In the case of long-term predictions, the empirical verification (i.e., against measured data after realization of the situation for which the predictions were made) is often impossible within a reasonable time. Hindcasts may help, but they are not always appropriate, since the model usually concerns future situations which are different from the past (otherwise, the model could just as well be replaced by a geostatistical extrapolation technique). Validation therefore depends to a large extent on other activities than empirical verification.

The possibility of empirical verification via hindcasts depends to a large extent on the quality of the available data. This goes for the wave, current, and sediment transport fields and for the morphological evolutions. Process data should provide information of the wave, current, or sediment transport *patterns* on the complex inlet topography. Due to the strong spatial variability, isolated point values are of little use. Remote sensing techniques can possibly offer new perspectives here, and also for measurements under rough conditions, when it is impossible to send out vessels. Depending on the timescale of interest, it may be more useful to have a quick RS-scan of the bed topography every month than to vary accurate depth soundings every five years.

6. Example of an ISE-Model Application

Problem Definition

The northern tip of the isle of Texel, in the Dutch Wadden Sea (Fig. 7), is subject to an erosion of some $5\ 10^5$ m^3/yr. So far, the sand deficit has been replenished via beach nourishments, which relieves the symptoms of the problem, but removes its causes. A historical reconstruction of the inlet evolution (Ribberink and De Vroeg, 1991) revealed that the sediment budget of the island tip is determined to a large extent by the course of the main ebb-channel, which used to change more or less continually between NE-SW and SE-NW. When the main ebb-channel had a NE-SW orientation, large amounts of sediment were deposited in front of the Texel coast, whereas this coast has a tendency to be sediment-starved when the channel had a SE-NW orientation. In the last 50 years, various engineering works have been made to protect the tip of the island, the main ebb-channel has stopped changing direction and has kept a SE-NW orientation. As a consequence, the isle of Texel keeps on eroding, and large amounts of sediment are continuously deposited on the vast beach plane which forms the tip of the next island, Vlieland.

As an alternative to an everlasting beach nourishment scheme, a long groyne near the tip of Texel was proposed (Fig. 8). This groyne would trap the north-going longshore drift and thus cause accretion of the coast at its south side. The corresponding erosion at the north side could be dealt with by connecting the groyne with the exisiting revetment around the northern-most tip of the island. Delft Hydraulics has carried out an extensive mathematical model study (Ribberink and De Vroeg, 1992) to investigate the effectiveness of various forms and locations of the groyne, the pattern and depth of scour around the tip, and the nature and extent of the effects on the inlet system.

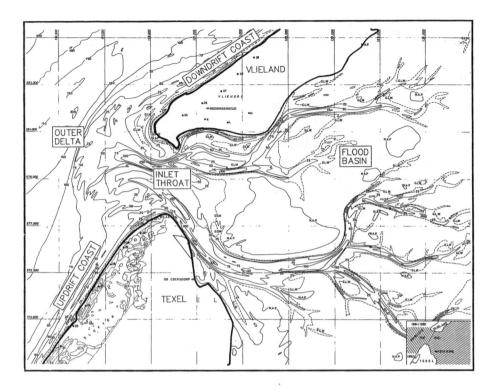

Fig. 7. Texel case: situation.

Model Composition

In order to predict the response of the inlet system to the cut-off of a sediment input of $5 \ 10^5 \ m^3/yr$, two-dimensional horizontal models are needed which describe the wave and current fields on a complex topography, including flooding and drying areas.

The wave field is assumed to be in equilibrium with the topography at any moment during the tide. The wave model has to include depth-refraction, shoaling, and breaking. Since the spatial variations in wave height are strong, but not abrupt, diffraction is probably less important. Because of the strong spatial variations of the current velocity, current refraction has to be included. Wind-induced growth and friction-induced decay are included in view of the large model area. In order to include some natural smoothing of wave-height

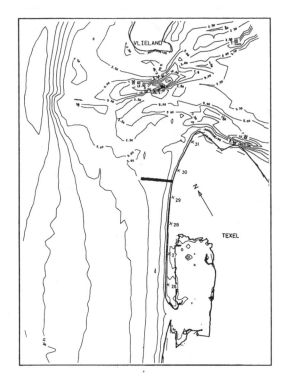

Fig. 8. Texel case: one of the proposed groyne alternatives.

variations, directional spreading is taken into account. The wave model which was used in this study is based on the HISWA package (Holthuysen et al., 1989) and meets these requirements.

This wave model is run at 20 tidal stages, which are chosen such that they provide maximum information on the time-distribution of the wave properties. The water level and the tidal current velocity at each stage are derived from a larger-scale tidal model which covers the entire Dutch Wadden Sea and part of the North Sea, and is run without taking wave effects into account.

The current field is computed with a 2D depth-integrated current model based on the TRISULA package (Stelling and Leendertse, 1992). It is basically a shallow-water tidal model, extended with (see Eqs. (1) through (3))

- the wave-induced effective stresses computed from the rate of wave-energy dissipation (Dingemans et al., 1987) via

$$F = D\frac{k}{\omega} \qquad (8)$$

in which

F = wave-induced effective stress vector [N/m^2],
D = wave-energy dissipation rate [N/ms],
k = wave number vector [m^{-1}],
ω = peak frequency of the wave field [s^{-1}],

- wave-induced enhancement of the bed shear stress, according to a parameterized version of Bijker's (1966) bed shear stress model with $\xi = 1$ (cf., Soulsby et al., 1993).

The vertical structure of the wave-induced current (mass-flux compensation, undertow, boundary layer streaming), as well as the mass flux in the depth-averaged equation of continuity, is ignored because cross-shore transport mechanisms are also ignored. A first estimate of the importance of curvature-induced secondary flows led to the conclusion that these can be left out of consideration.

Estimates of the relevant sediment transport properties revealed that it must be possible to use a transport formula here. This formula should apply to situations with combined waves and currents. For reasons of feasibility, "cross-shore" transport mechanisms due to wave asymmetry, undertow, streaming, etc., are left out of consideration. The remaining current-borne transport (the waves and current stir the sediment, the current transports it) is modeled with Bijker's (1971) formula. This formula has to be applied in each run at each of the 20 tidal stages and at each point of the computational grid. Clearly, this requires an efficient algorithm, which is achieved by parameterizing the formula.

Model Domain

The model domain is outlined in Fig. 9(a). Inside the basin, the boundary follows the watershed, which encloses an area of nearly 200 km^2. The island coasts form closed boundaries, which extend alongshore until "far away" from the inlet. This longshore extent is generally chosen such that the expected changes in the inlet are not felt at the open boundaries. This means that the boundary conditions can be the same before and after construction of the

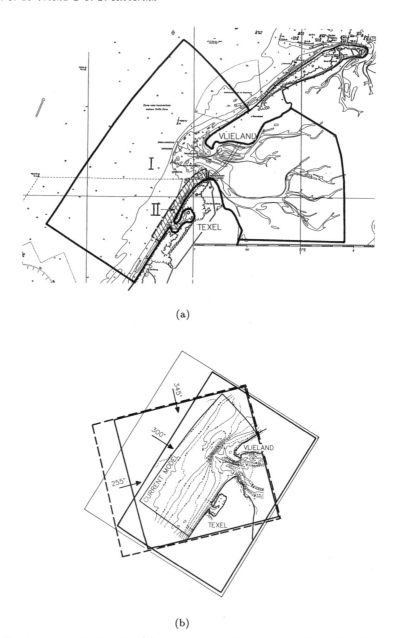

Fig. 9. Texel case: model domain, (a) current and sediment-transport models, (b) wave models.

groyne. As a consequence, the seaward part of the model covers an area of some 240 km^2.

The version of the HISWA model which is utilized here works on a rectangular grid of which the orientation of the grid lines in one direction coincides more or less with the direction of wave incidence. Since various angles of wave incidence are investigated in this study, there are various wave model grids, each of which encompasses the current-model domain (Fig. 9(b)). In principle, the model works with three grids for each run, viz. an input grid on which the bed topography is given, a computational grid, and an output grid on which the output parameters are provided. Although the interpolations are fully automized, the variety of model domains requires extra administration, and thus it is a source of errors. Later applications of this model therefore work on the same domain as the current model.

Geometrical Schematization

The wave computations are made on a rectangular grid, but the output can be given on any arbitrary set of points. Therefore, the sediment transport computations are made on the boundary-fitted curvilinear grid of the current model (Fig. 10). This grid contains 14,157 (117 × 121) points, out of which some 4200 are permanently inactive because they are located on land. The grid size varies in space, with a minimum size of 40 m by 125 m in the area of interest. Criteria for the grid generation are orthogonality (required by the current model), smoothness, and spatial resolution.

Resolution-demanding parts of the model are the main tidal channels and the surf zones. Via the CFL criterion, the small grid size in the surf zones leads to small time steps (30 seconds, which corresponds with a CFL number of a maximum of 15).

The bed topography, as represented by the bathymetric maps available, was translated to the curvilinear grid using a discretizer. Another possibility would be a digital terrain model, provided that the criteria for grid and topography optimization are adjusted to the demands of a tidal model.

Boundary Conditions

The boundary conditions for the wave model concern the wave height and direction at the upwind boundary. The wave energy is spread over the spectral directions according to a $\cos^2$ function. No boundary conditions are imposed

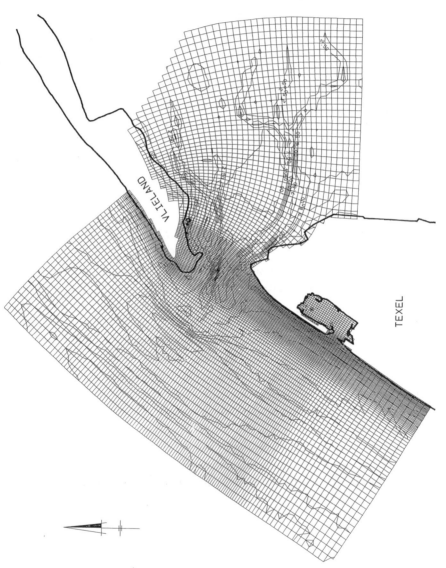

Fig. 10. Texel case: computational grid of the current model.

at the lateral boundaries, which are assumed to be more or less parallel to the direction of wave propagation. Thus, they are treated as wave rays, across which there is no net energy flux. All energy is supposed to be dissipated before the waves reach the landward boundaries, so there is no reflection and there is no need to impose boundary conditions there. This is consistent with the refraction approximation which underlies the model.

Tidal current models of this size are driven by the conditions imposed at the open boundaries. The present model contains basically two open boundaries, one behind the gorge and one in the open sea. At the basin boundary and the NE part of the seaward boundary, the water-level variation is prescribed, and at the SW and NW parts of the seaward boundary the tidal current velocity is imposed.

The wave-induced current is not driven at the boundaries, but by the effective wave-induced stress, F, inside the domain. This means that this part of the current should be disturbed as little as possible by the boundary conditions. In practice, this means that a cross-shore boundary which intersects an area with a strong wave-induced current or set-up is usually treated as an axis of symmetry with zero normal derivatives. The wave-induced contribution to the velocity or the water-level distribution along such a boundary can be computed from a longshore uniform current or set-up model.

Representation of Natural Conditions

The model input has to be chosen such that a reasonable estimate is obtained for the long-term residual transport field. Unfortunately, the relative importance of the transport-generating mechanisms varies through the model area, so an input schematization which is adequate at one point of the model domain is not necessarily adequate at all other points.

In order to tackle this unresolved problem, the pragmatic "multiple representative wave" approach (cf., De Vriend et al., 1993a) is used. This leads to a set of representative wave conditions which are shown in Table 1. The method also yields a representative tide, with an amplitude some 10% above the mean. A tide of this amplitude which had already been simulated with the ambient tidal model for the entire Wadden Sea is the astronomical tide of 21 March 1976. Therefore, this tide was chosen as morphologically representative of all tides.

For each situation, viz. without groyne or with one of the alternative forms of groyne, the number of wave-model runs is sixty (three sectors × twenty tidal

Table 1. Representative wave conditions.

Name	Compass angle	H_s	T_p	Occurrence (%)
Westerly	255°	1.48 m	6.7 s	39.2
Northerly	345°	1.33 m	6.6 s	21.4
Storm	300°	4.60 m	10.0 s	1.0
Tide only	–	–	–	38.4

stages), and the number of current-model runs is four (tide only and three wave conditions).

Parameter Setting and Validation

Some of the model parameters are chosen on the basis of professional judgement, others are calibrated against known properties of the system. Examples of the former kind are the coefficients in the wave-energy dissipation model, the bottom friction factor in the wave model, Manning's n (0.027), and the eddy viscosity (2 m²/sec) in the current model. Note that the latter value has little to do with the physical eddy viscosity: it mainly accounts for effects of the horizontal discretization.

Important elements of validation are the long-term residual total transport of sediment through the inlet gorge, and the gap of $5\ 10^5$ m³/yr in the nearshore sediment balance near the tip of the island. Other validation material is provided by a coastline model which is run with a very detailed representation of the wave climate, and by additional experiments with the present model with a detailed representation of the wave climate (Negen, 1994). These experiments made clear that the wave schematization technique yields reasonably good estimates of the residual transport throughout the model domain.

In order to have the long-term residual transport through the gorge right, the tidal asymmetry effect had to be enhanced by amplifying the flood transport and reducing the ebb-transport, both by 20%. The computed longshore drift is somewhat smaller than that computed with the coastline model. This difference is attributable to the wave-climate schematization, and it is compensated by enhancing the weight of the transport due to westerly winds by a factor of 1.7 and that due to northerly winds by a factor of 2.6.[a]

---

[a] These factors are not alarming, since the original weights are proportional to the percentage of occurrence of the wave height. In view of the the nonlinearity of the transport formula, this proportionality is not quite realistic.

Large-Scale Processes Without Groyne

The first model application concerns the analysis of the hydrodynamic and sediment transport processes at work around the inlet. Figure 11 shows a typical result based on the weighted sum of results from the various model runs. These results are interpreted in terms of sediment drift patterns at an essentially higher aggregation level than the individual transport vectors.

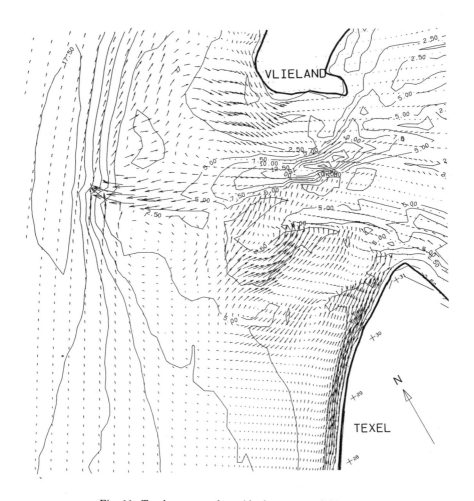

Fig. 11. Texel case: yearly residual transport field.

Figure 11 shows a pronounced ebb-dominated transport in the main ebb-channel. The sediment is deposited at the edge of the outer delta, from where it is removed under storm conditions. Furthermore, there are distinct flood-dominated transports near the island tips; in the case of Texel, it is somewhat offshore. Between this flood "channel" and the coast, there is a weakly developed ebb-channel and a pronounced longshore drift into the inlet.

One reason why this longshore drift is so pronounced is due to the orientation of the coastline, which is by no means perpendicular to the predominant incoming wave direction. Another reason is the presence of the outer delta, which shelters the island tip from high northerly waves. Thus, the westerly waves and the associated NE-bound longshore drift are predominant in that area. Once the drift enters the inlet, the tidal current takes over and transports the sediment via the ebb-dominated main channel towards the edge of the delta and from there mainly to the tip of the Vlieland island.

The storm-induced transport pattern is quite different from the weighted average one. The representative wave heights in each sector are meant to yield a good approximation of the longshore drift at a straight, exposed coast. The result is a rather moderate wave height (see Table 1), and waves which only break on the beaches and on the shoals of the outer delta. Storm waves, however, break on the delta edge and drive a strong onshore and longshore transport there (Fig. 12). Clearly, these storm waves contribute significantly to the onshore "bulldozer effect", which counteracts the tendency of the tide to extend the delta further seawards.

Large-Scale Groyne Effects

A long groyne perpendicular to the shore will interrupt the longshore drift. As a consequence, sediment will pile up at the updrift side, and erosion will occur at the downdrift side. At a fully exposed coast without a tide, this process would continue until the coastline has built out so far that the amount of sediment which bypasses the groyne is equal to the undisturbed longshore drift. In the present situation, the tidal current also moves around the tip of the breakwater and forms another bypass mechanism. As a consequence, the coastline is unlikely to build out too far. An ISE model is unable to predict the extent of this process, but it gives an indication of the bypass potential of the tidal current. Figure 13 shows that this potential is significantly less than the longshore drift. This means that a substantial build-out of the coastline can be expected.

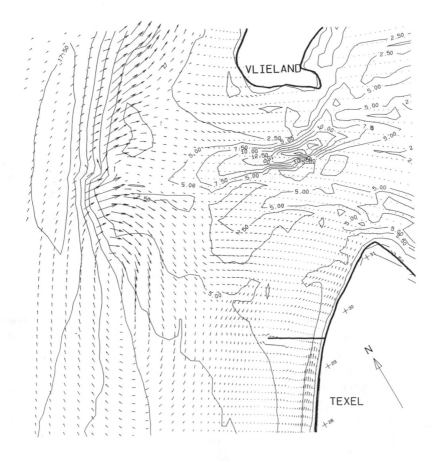

Fig. 12. Texel case: storm-induced transport field.

This is corroborated by the accretion/erosion patterns in Fig. 14. Figures 13 and 14 also show that at the downdrift side the longshore transport picks up again, which leads to severe erosion. This effect is stronger the further the groyne lies from the island tip.

At the scale of the inlet as a whole, Fig. 14 shows that the effects of the groyne are not restricted to its immediate vicinity, but that they extend all over the inlet. The nature of the effects does not become very clear, because the figure shows a sequence of erosion and deposition areas.

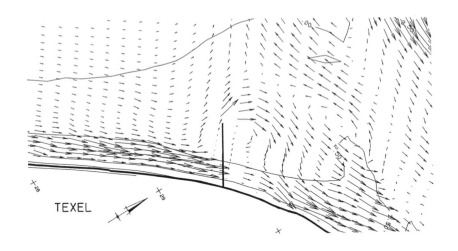

Fig. 13. Texel case: transport around the groyne.

The latter phenomenon is typical for ISE models. Morphological features have a strong tendency to migrate, but since ISE models lack the morphodynamic interaction, they only show the onset of this migration, which is a coupled pair of erosion and deposition areas (Fig. 15). Moreover, experience with morphodynamic model applications shows that, initially, the model tends to remove irregularities from the given initial bed topography. Thus, the topography is moulded into a generally smoother form which suits the model. From then on, a much slower and physically more realistic dynamic process takes place (cf., De Vriend et al., 1993b). Clearly, ISE models represent the initial moulding process.

Local Groyne Effects

The principal near-field effect of the groyne is the formation of a scour hole near its tip. Again, the ISE model results ought to be interpreted with care here. Instead of a single scour hole, Fig. 14 shows a pair of erosion and deposition areas, with the transition exactly in front of the groyne. This might lead to the erroneous conclusion that the most severe erosion occurs at some distance from the tip, and thus an underestimation of its threat to the structural stability.

Mathematical Modeling of Meso-Tidal Barrier Island Coasts 179

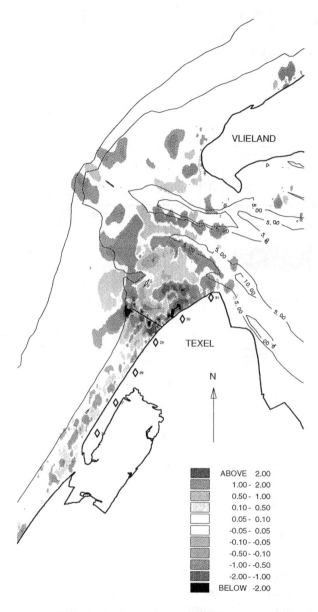

Fig. 14. Texel case: large-scale ISE pattern.

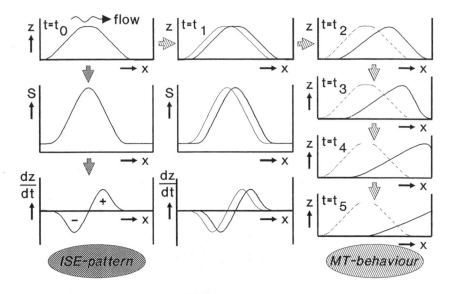

Fig. 15. Migration of morphological features in ISE models and MTM models.

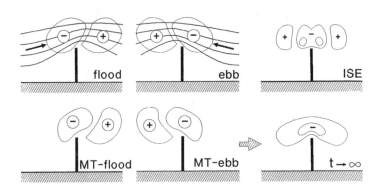

Fig. 16. Morphological evolution around a groyne.

Figure 16 explains what actually happens. Initial erosion occurs where the flow accelerates, deposition where it decelerates. In the longer run, however, the topography tends towards a state where the sediment flux through a streamlane is constant. In this situation, the scour hole coincides more or less with the area of maximum velocity.

7. ISE Modeling Practice

The concept of ISE models is rather simple, at least if adequate wave, current, and sediment transport solvers are available. Yet, the above example shows that their practical operation involves points of attention, such as

- software infrastructure (steering programme, data transfer system, pre- and postprocessing facilities, etc.). Efficiency and structure are important issues here, because many runs have to be made in order to cope with the natural variability of conditions.
- model composition, i.e., which combination of process models should be applied in order to be able to achieve the objectives of the model study. ISE models pose specific problems here, because a limited number of "snapshot" runs has to provide sufficient information on the morphological evolution to be expected. On the other hand, unnecessary refinements lead to inefficiency and sometimes even to interpretation problems.
- geometrical schematization (model boundaries, grid definition, topographical schematization, etc.). Many choices and trade-offs have to be made here. One example is grid resolution versus computer costs.
- input schematization, in order to have a manageable number of runs representing the whole spectrum of possible combinations of relevant input conditions (mean water level, deep-water wave height/period/direction, tidal constituents, etc.). This schematization is closely linked to the number of model runs to be made and to the derivation of quantitative predictions from the results of these runs.
- data transformation, if the various modules operate on different grids and in different domains (cf., the wave models in the above example). Unless special techniques are applied, data transformation gives rise to a loss of accuracy.
- interpretation of results. This is a particularly challenging part of ISE modeling, as may have become clear from the above example (also see Sec. 9).

In practice, the main use of ISE models lies in the analysis of residual tansport patterns under various conditions, and in the prediction of their reponse to large disturbances, such as engineering works, land reclamation, sand mining, subsidence due to oil/gas mining, etc. ISE models are not suited to analyze the subtle balance of processes in an inlet system in (near-)equilibrium.

8. Specific Aspects of MTM-Modeling

MTM models of tidal inlets will often be used to predict the effects of accelerated mean sea-level rise, enhanced subsidence due to oil and gas mining, land reclamation, sand mining, coastal nourishments, engineering structures, etc. Many of these disturbances boil down to source or sink terms in the sediment balance equation

$$(1-\varepsilon_p)\frac{\partial z}{\partial t} + \frac{\partial S_x}{\partial x} + \frac{\partial S_y}{\partial y} = -\frac{dMSL}{dt} - \frac{dz_{ref}}{dt} + \Phi_{dep} - \Phi_{er} \qquad (9)$$

in which

MSL = mean sea level (m/yr),
z_{ref} = reference bed level (subsiding) (m),
Φ_{er} = deposition flux (e.g., nourishment) (m^3/m^2/yr),
Φ_{dep} = ersoion flux (e.g., mining) (m^3/m^2/yr).

These terms give rise to a forced response of the system, on top of the inherent behavior which is present anyway.

The numerical integration of the sediment balance equation may seem to be simple and straightforward, and so it is if the transport field is known. In an MTM model, however, the equation is included in a time-loop, due to which there is a feedback of z on S. As a consequence, the numerical problem becomes much more complicated and instability-prone. One-step explicit schemes are the most convenient in this case, and schemes of the Lax-Wendroff type turn out to be the most robust (cf., De Vriend et al., 1993b; also see Peltier et al., 1991). Like other explicit schemes, they involve a CFL-type stability criterion.

As the tidal computation is usually the most time-consuming, minimization of the number of its calls is an effective way to reduce computer costs. One rather robust technique to achieve this is the "continuity-update" (see Fig. 18 for its incorporation in the flow chart), which reads in its simplest form

$$(\bar{u}h)_{t+\Delta t} = (\bar{u}h)_t . \qquad (10)$$

More sophisticated versions make use of the full equation of continuity. Tests with horizontally 2D morphological models have shown that very significant savings of computer time can be achieved with this technique (Latteux and Peltier, 1991; De Vriend et al., 1993b).

The above technique does not include the momentum equations, so it does not take the dynamic interaction between the current and the bed topography into account. Thus, the propagation character of morphological changes is ignored. This introduces systematic errors in the solution, which increase with the number of consecutive updates with this technique. However, analyses have shown that these errors tend to disappear as soon as the full hydrodynamic model is applied again. Thus, the intermittent as well as long-term application of the full model and a number of continuity-updates yields acceptable results.

9. Examples of MTM Model Applications

Texel Case

The model system described in the foregoing has recently been extended to a morphodynamic simulation system. In essence, this boils down to feeding the bed level changes from the sediment balance equation back into the hydrodynamic and sediment modules (Fig. 1(b)). This creates a time-loop which simulates the mesoscale morphodynamic behavior of the inlet system.

Negen (1994) describes the application of this simulation technique to the Texel case, be it with the groyne at a slightly different location. Figure 17 shows a typical result, after a year's simulation, continuously with westerly waves (Table 1). The results generally confirm the effect of the groyne on the coastline position. Striking differences from the ISE model results are:

- the formation of a channel along the downdrift side of the groyne,
- the deepening of the weakly developed NE-SW ebb-channel near the tip of the groyne, and shoaling of the main ebb-channel where this crosses the course of the NE-SW channel,
- the onshore migration of the shoals on the outer delta.

The first point indicates that further study is needed to design a stable groyne. The second finding may indicate that the groyne tends to promote the reorientation of the main ebb channel, back to the NE-SW position which is favorable for the sediment budget of Texel. The third observation shows that the present simulation probably does not represent reality, because the mechanisms which tend to move the shoals offshore are underestimated.

In summary, this application shows the potential of fully dynamic simulations of tidal inlet morphodynamics. On the other hand, the present case definitely requires further attention before definitive conclusions can be drawn from the simulation.

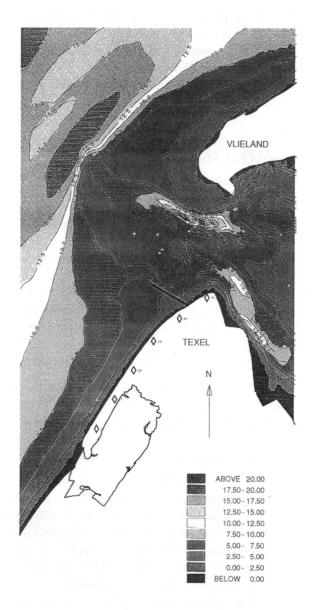

Fig. 17. Texel case: morphological evolution after one year, according to an MTM simulation with westerly waves. (Negen, 1994; reproduced by courtesy of Delft Hydraulics)

Frisian Inlet Case

Another less complicated application of a morphodynamic simulation model concerns the Frisian Inlet (Wang et al., 1991, 1995; see Sec. 5 of the previous chapter for a description of the situation). The objectives of this study were

- to hindcast the system's response to the closure of the Lauwerszee,
- to assess residual transport patterns, and
- to investigate channel migration mechanisms,

especially in the basin and the inner part of the gorge. This explains why this study could be undertaken with a model which does not include waves.

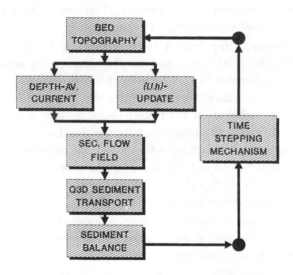

Fig. 18. Frisian Inlet morphodynamic simulation model: flow chart.

Figure 18 outlines the flow chart of this model, including a "continuity-update" technique to save computer time (see Sec. 8). The tidal module is essentially the same as in the Texel case, but extended to $2\frac{1}{2}$D in order to take curvature- and coriolis-induced secondary flows into account. The sediment transport module is based on a $2\frac{1}{2}$D advection/diffusion model for the concentration, with the equilibrium concentration derived from Van Rijn's (1984) transport formula (although other options are open). The choice of these modules was made on the basis of a number of orientating computations.

The model domain covers not only the area of interest, but also the outer delta and part of the open sea, in order to have the seaward boundary at a sufficient distance away from the dynamic area to allow for the same tidal boundary conditions throughout the simulation.

The schematization of the inlet geometry turned out to be critical for the system's morphodynamic stability. If the emerging shoal in the center of the inlet, the Engelsmanplaat, is omitted, the system becomes unstable. One might claim that the model should be able to reproduce the formation of this feature if that is absent from the initial state. Closer investigation of the geological structure of the Engelsmanplaat (Oost and De Haas, 1993), however, revealed that it consists mainly of highly resistant pleistocene clay, i.e., it is not the result of a recent morphodynamic process. This feature therefore has to be included in the model as a permanent element.

As a consequence, the Frisian Inlet becomes a double inlet, with the Pinkegat subsystem in the west and the Zoutkamperlaag subsystem in the east. The large difference in size between these subsystems is reflected in a large difference in the morphodynamic timescale. The model's morphological time step has to be based on the fastest evolving subsystem, the Pinkegat. Thus, it is much smaller than necessary for the Zoutkamperlaag. Within the amount of computer time allotted to this model, the response of the Zoutkamperlaag could therefore be only simulated if the topography of the Pinkegat area is fixed. Clearly, such a drastic measure is bound to cause problems at the transition to the mobile-bed part of the model.

The seaward boundary conditions (tide, sediment transport) form another critical element. Disturbances due to errors or inconsistencies in these conditions are translated into bed-level perturbations which tend to migrate into the model domain and sometimes even blow up. In general, a morphodynamic simulation requires one boundary condition, either to the transport or to the bed level, at every inflow boundary. This poses a special problem to tidal models, in which the open boundaries keep swapping between the inflow and the outflow type. This problem is largely unresolved as yet and definitely requires further attention.

The model results obtained so far concern mainly the shorter-term phenomenon of channel formation and migration, especially in the Pinkegat area. The model, as well as field observations (Oost and De Haas, 1993; Oost, 1995) show a cyclic behavior of the channel pattern (Fig. 19), though at different timescales. This discrepancy is probably due to the absence of waves from

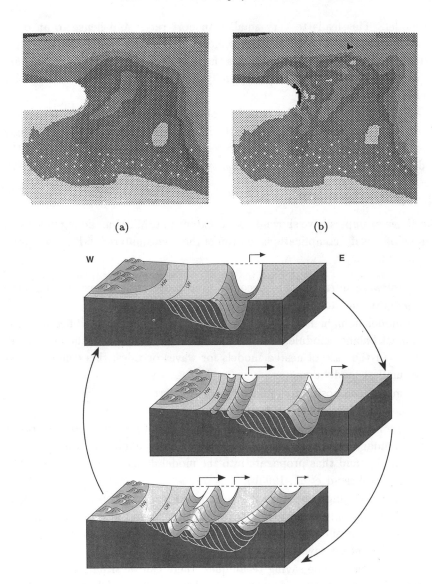

Fig. 19. Frisian Inlet: channel pattern evolution, (a) initial condition in the model, (b) predicted evolution, (c) observed cyclic evolution. (courtesy of Z. B. Wang (a and b) and A. P. Oost (c))

the model. This leads to very small transport rates, and hence to an overprediction of the morphological timescale. Yet, the character of the migration pattern is reproduced, which indicates that the model contains the right channel migration mechanism. Further analysis should reveal the details of this mechanism.

Another indication of the model's potential concerning channel formation follows from an experiment in which the initial bed topography was a strongly smoothed representation of the measured topography. It turns out that the model tends to restore a channel/shoal system of a similar nature as before.

10. MTM Modeling Practice

The above examples have shown to some extent that MTM modeling involves a number of specific complications, on top of those encountered in ISE modeling. In brief summary:

- The software infrastructure becomes even more critical now, because the coupling of the modules has to be completely automized.
- The model domain has to be the same for the process modules and the sediment balance module, in order to avoid discontinuities in the bed topography. In the case of nested models for waves or tides, this concerns only the smallest model domains.
- The model composition should be such that errors cannot build up to an unacceptable level during the simulation.
- The boundary conditions need to be defined with care. In general, errors in the boundary conditions tend to be reflected in an amplified form in the bed topography, and thus propagate into the model domain (Wang et al., 1991; also see De Vriend et al., 1993b).
- The time-stepping procedure in the morphodynamic loop deserves special attention, not only because it is instability-prone, but also because there is a large amount of computer time involved, whence clever time stepping may lead to significant savings.
- In spite of these time-saving techniques, there can be situations where the morphological timescales vary strongly through the model area (cf., Wang et al., 1991; 1995). In that case, the shortest timescale determines the maximum allowable morphological time step, which may imply that large amounts of computer time are needed to cover the largest timescale of interest.

- Since the simulation covers a certain period of time, it is possible to mimic the natural variation of conditions via input scenarios for waves, water levels, tides, and surges. Such a scenario is only one realization of a stochastic process, so the simulation should be repeated a number of times in order to get a feel for the variability of the output.
- The same goes for verification against field observations, which are the result of only one realization of the input conditions. Therefore, such a verification cannot validate the variation of the output due to different input scenarios.

In general, it is very important to clearly identify the objectives of the model study. All-purpose simulation models are not yet available, so the model composition and set-up have to be derived from the objectives of the study, via an analysis of the relevant phenomena.

MTM models, especially those including waves, are still at an early stage of development. Now that the numerical and software problems have largely been overcome, the focus of attention should return to the physical processes and their interactions, including the equilibrium state to which the system tends. Key issues of research are the tenability of the 2D depth-integrated model concept and its $2\frac{1}{2}$D or quasi-3D alternatives, and ways to deal with the natural variability of the input conditions.

11. Process-Based Long-Term Models

One way to achieve a long-term morphological model concept for tidal inlets is to formally average the constituting mathematical-physical equations over the space and timescales which are smaller than those of the phenomena of interest. Since these constituting equations are nonlinear, this involves closure terms which have to be modeled in terms of the large-scale dependent variables (cf., the Reynolds-stress terms in turbulent flow models, and the radiation-stress terms in wave-driven current models). This can be done empirically, or based on a theoretical analysis of the relevant interaction processes.

Krol (1990) describes a method of formal averaging for a simple 1D morphodynamic model of a tidal estuary. This work has been followed up by various other investigators (Fokkink, 1992; Schuttelaars and De Swart, 1994), and recently in two dimensions (Schuttelaars, 1994, private communication).

A simple example of formal averaging concerns a narrow basin of length L and width B (cf., Van Dongeren and De Vriend, 1994). L is much smaller than the tidal wave length, and B is constant, whence it can be taken equal to 1 without loss of generality. The constituting equations are:

(1) the equation of continuity of the tidal motion

$$\frac{\partial \zeta}{\partial t} + \frac{\partial[(\zeta - h)\,U]}{\partial x} = 0, \qquad (11)$$

in which

ζ = water surface elevation above MSL,
h = bed level with respect to MSL (so usually $h < 0$),
U = cross-sectional average velocity,
x = spatial coordinate along the basin.

(2) the dynamic equation for the tide, which reduces to

$$\frac{\partial \zeta}{\partial x} = 0, \qquad (12)$$

(3) the sediment transport formula

$$S(x,t) = a\,U^b, \qquad (13)$$

in which a and b are constants.

(4) the sediment balance equation

$$\frac{\partial h}{\partial t} + \frac{\partial S}{\partial x} = 0. \qquad (14)$$

Integration of Eq. (11) over the length of the basin yields for $\zeta \ll h$.[b]

$$U(x,t) \approx \frac{L - x}{-h(x,\tau)}\frac{d\zeta}{dt} \qquad (15)$$

in which ζ denotes a time coordinate at the slow timescale of the morphological changes. With Eq. (13) and upon averaging over the tidal cycle yields

$$\bar{S}(x,\tau) = a\left[\frac{L-x}{-h(x,\tau)}\right]^b Z_T \quad \text{in which} \quad Z_T = \frac{1}{T}\int_0^T \left(\frac{d\zeta}{dt}\right)^b dt. \qquad (16)$$

[b]This assumption is made for explanatory reasons. Clearly, it does not hold in the shallow parts of the basin. Yet the resulting linear distribution of $h_e(x)$ turns out to be a reasonable approximation (Schuttelaars and De Swart, 1994).

Formal averaging of the sediment balance equation (14) over the tidal cycle yields

$$\frac{\partial h}{\partial \tau} + \frac{\partial \bar{S}}{\partial x} = 0 \qquad (17)$$

or, with Eq. (16),

$$\frac{\partial h}{\partial t} + \frac{b\bar{S}}{-h(x,\tau)} \frac{\partial h}{\partial x} = \frac{b\bar{S}}{L-x}. \qquad (18)$$

This equation describes the evolution of $h(x, \tau)$ at the morphological timescale. Apparently, this equation is of the forced nonlinear advection type, with the advection speed bS/h, which is found in many other morphological systems with a transport formula of the same type as in Eq. (13). Note that, for $b > 1$, the source term vanishes towards the back end of the basin. This means that the evolution process proceeds ever slower as x approaches L.

The equilibrium state which follows from Eq. (18) is given by

$$\frac{\partial h_e}{\partial x} = -\frac{h_e}{L-x} \quad \text{whence} \quad h_e \sim L - x. \qquad (19)$$

As long as no part of the basin falls dry at low water, the tidal prism at any point x is also proportional to $L - x$. Hence, the equilibrium cross-sectional area in this case is proportional to the tidal prism. This corresponds with O'Brien's (1931, 1969) empirical findings.

Figure 20 shows the result of a fully nonlinear analytical model of the morphological evolution in a dead-end tidal channel without flats (Schuttelaars and De Swart, 1994). It shows that the above simplified linear model is a reasonable approximation. Note that the evolution process becomes infinitely slow towards the end of the channel, because at that point, the current velocity and the transport rate go to zero (also see Van Dongeren and De Vriend, 1994).

The derivation of the above linearized model shows the essence of many long-term morphological model concepts, viz. to express the residual transport, and hence, the time-derivative of the bed level, in terms of the bed elevation. The result is an evolution equation in h. The same approach is taken by Bakker (1995) to derive a bed evolution equation for a tidal channel in a network, i.e., with two open ends. Bakker and De Vriend (1995) use the result in a network model of a near-resonant basin. One of the research questions which they address is to what extent a small interference at one place in the basin can have large effects elsewhere.

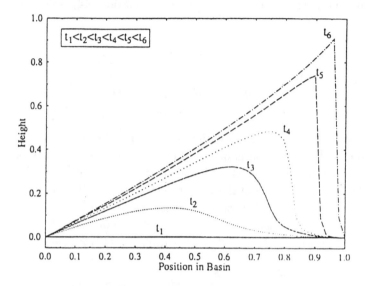

Fig. 20. Morphological evolution of a dead-end tidal channel.

This theoretical approach of tidal inlet morphodynamics is quite complex on the one hand, and quite powerful on the other. Analyses of the mathematical system, once it has been formulated, are able reveal the inherent morphodynamic behavior, such as channel pattern formation and stability of channel/shoal systems. This information can not only be of use for the validation of numerical models, it may even indicate the extent to which numerical models are able to handle these nonlinear phenomena.

12. Conclusions

Computer-intensive multidimensional model types are now available for application to tidal inlets. ISE model systems have reached a more or less operational stage (Latteux and Peltier, 1992; Broker Hedegaard et al., 1993; Roelvink et al., 1994; Chesher et al., 1993) and have proven their use to describe physical processes, such as residual transport patterns, under a variety of input conditions. This information can be used to analyze how waves, tide, currents, and sediment transport influence each other in an area as complex as a tidal inlet, and to estimate the impact of human interference in the system (engineering structures, land reclamation, sand mining, etc.).

Process-based morphodynamic simulation models are still in an early stage of development. It is technically possible to run such models and pilot applications are being made, but the stage of routine practical application has not yet been reached. This requires further investment into a more thorough understanding of the processes and mechanisms at work, and a further build-up of know-how, experience, and confidence in these models.

One objective of further research should be a better insight into the relevance of phenomena at the process level to larger-scale morphological behavior in a given situation. Thus, it must become clearer what physics should be included in the constituent models in order to simulate these evolutions.

Another research issue is how to deal with the natural variability of the conditions which drive a tidal inlet model, and how to interpret predictions which result from an assumed future input scenario. Conversely, it should become clear which aspects of the input time series are important to the morphological evolution of the system.

Finally, the formal averaging approach and the corresponding theoretical analyses should be pursued further, if it were only for the large amount of insight which they would provide.

At the development level, knowledge and tools for the adequate selection of model components would be welcomed. Also, the insight into the accuracy and reliability of these compound dynamic models should be improved, in conjunction with practical methods to deal with the limited predictability of inputs.

Acknowledgements

Many of the examples in this chapter stem from projects which were sponsored by The Netherlands Ministry of Transport and Public Work (Rijkswaterstaat), under the heading of design projects (Eierlandse Gat), or under the Coastal Genesis research programme (Frisian Inlet).

The authors gratefully acknowledge Dr. Z. B. Wang of Delft Hydraulics and Mr. A. P. Oost of Utrecht University, who made unpublished material available.

References

Bailard, J. A. (1981). An energetics total load sediment transport model for a plane sloping beach. *J. Geoph. Res.* **86**(C11): 10938–10954.

Bakker, W. T. (1995). Effect of tidal resonance on the morphology of Wadden and estuaries (In preparation).

Bakker, W. T. and H. J. de Vriend (1995). Resonance and morphological stability of tidal basins. *Marine Geology.* **126**: 5–18.

Bakker, W. T. and D. S. Joustra (1970). The history of the Dutch coast in the last century. *Proc. 12th Int. Conf. Coastal Engineering.* ASCE. 709–728.

Berlamont, J., M. Ockenden, E. Toorman and J. Winterwerp (1993). The characteristics of cohesive sediment properties. *Coastal Eng.* **21**(1-3): 105–128.

Bijker, E. W. (1966). The increase of bed shear in a current due to wave action. *Proc. 10th ICCE*, Tokyo, Japan. 746–765.

Bijker, E. W. (1971). Longshore transport computations. *J. Wtrwys., Harb. and Coast. Eng.* **97**(WW4): 687–701.

Brøker Hedegaard, I., K. Mangor and M. J. Lintrup (1993). Modeling of sediment transport in connection with tidal inlets. *Proc. Comp. Appl. Coastal and Offshore Eng. (ICE-CACOE '93)*, Kuala Lumpur, Malaysia.

Chesher, T. J., H. M. Wallace, I. C. Meadowcroft and H. N. Southgate (1993). PISCES, a morphodynamic coastal area model; first annual report. HR Wallingford, Report SR 337.

De Vriend, H. J. (1987). 2DH mathematical modeling of morphological evolutions in shallow water. *Coastal Eng.* **11**(1): 1–27.

De Vriend, H. J. (1994). Two-dimensional horizontal and weakly three-dimensional models of sediment transport due to waves and currents. In *Coastal, Estuarial and Harbour Engineers' Reference Book*, eds. M. B. Abbott and W. A. Price, E & FN Spon. 215–238.

De Vriend, H. J. and J. S. Ribberink (1988). A quasi-3D mathematical model of coastal morphology. *Coastal Engineering 1988 Proc.*, ed. B. L. Edge. ASCE. 1689–1703.

De Vriend, H. J. and M. J. F. Stive (1987). Quasi-3D modeling of nearshore currents. *Coast. Eng.* **11**(5&6): 565–601.

De Vriend, H. J., W. T. Bakker and D. P. Bilse (1994). A morphological behavior model for the outer delta of mixed-energy tidal inlets. *Coast. Eng.* **23**(3&4): 305–327.

De Vriend, H. J., M. Capobianco, T. Chesher, H. E. de Swart, B. Latteux and M. J. F. Stive (1993a). Approaches to long-term modeling of coastal morphology: A review. *Coast. Eng.* **23**(1-3): 225–269.

De Vriend, H. J., T. Louters, F. Berben and R. C. Steijn (1989). Hybrid prediction of a sandy shoal evolution in a mesotidal estuary. *Hydraulic and Environmental Modelling in Coastal, Estuarine and River Waters*, eds. R. A. Falconer, P. Goodwin and R. G. S. Matthew. Gower Technical. 145–156.

De Vriend, H. J., J. Zyserman, J. Nicholson, Ph. Péchon, J. A. Roelvink and H. N. Southgate (1993b). Medium-term 2-DH coastal area modeling. *Coast. Eng.* **21**(1-3): 193–224.

Deigaard, R., J. Fredsøe and I. Brøker Hedegaard (1986). Suspended sediment in the surf zone. *J. Wtrwy., Port, Coast. and Oc. Eng. ASCE* **112**(1): 115–128.

Dingemans, M. W., A. C. Radder and H. J. de Vriend (1987). Computation of the driving forces of wave-induced currents. *Coast. Eng.* **11**(5&6): 539–563.

Dronkers, J. (1986). Tidal asymmetry and estuarine morphology. *Neth. J. Sea Res.* **20**: 117–131.

Fokkink, R.J. (1992). Fundamental considerations on morphodynamic modeling in tidal regions; Part II: A semi-analytical method for tidal basins. Delft Hydraulics, Research Report Z 331-II. 36.

Fredsøe, J. and R. Deigaard (1992). *Mechanics of Coastal Sediment Transport. Advanced Series on Ocean Engineering*, Vol. 3. World Scientific. 369.

Friedrichs, C. T. and D. G. Aubrey (1994). Tidal propagation in strongly convergent channels. *J. Geoph. Res.* **99**(C2): 3321–3336.

Fritsch, D., C. Teisson and B. Manoha (1989). Long-term simulation of suspended sediment transport, application to the Loire estuary. *Proc. 23rd IAHR Congress*, Ottawa, Canada. C276–284.

Holthuysen, L. H., N. Booij and T. H. C. Herbers (1989). A prediction model for stationary short-crested waves in shallow water with ambient currents. *Coast. Eng.* **13**: 23–54.

Horikawa, K. (1988). *Nearshore Dynamics and Coastal Processes.* University of Tokyo Press. 522.

Kalkwijk, J. P. Th. and R. Booij (1986). Adaptation of secondary flow in nearly horizontal flow. *J. Hydr. Res.* **24**(1): 19–37.

Katopodi, I. and J. S. Ribberink (1992). Quasi-3D modeling of suspended sediment transport by currents and waves. *Coast. Eng.* **18**: 83–110 (err. **19**:339).

Kirby, J. T., R. A. Dalrymple and H. Kaku (1994). Parabolic approximations for water waves in conformal coordinate systems. *Coast. Eng.* **23**: 185–213.

Krol, M. S. (1990). The method of averaging in partial differential equations. Ph.D. Thesis, Utrecht University. 81.

Latteux, B. and E. Peltier (1992). A two-dimensional finite element system of sediment transport and morphological evolution. *Int. Conf. on Comp. Mod. of Seas and Coastal Regions*, Southampton, UK.

Negen, E. H. (1994). Morphological study with 2DH numerical models near Eierland (Texel). Delft Hydraulics/Delft University, Technical Report H1887/H460. 68.

O'Brien, M. P. (1931). Estuary tidal prism related to entrance areas. *Civ. Eng.* **1**(8): 738–739.

O'Brien, M. P. (1969). Equilibrium flow areas of inlets on sandy coasts. *J. Wtrwy. Harb. ASCE* **95**(WW1): 43–52.

Oost, A. P. (1995). The cyclic development of the Pinkegat Inlet system and the Engelsmanplaat/Smeriggat, Dutch Wadden Sea, over the period 1832-1991. Dynamics and sedimentary developments of the Dutch Wadden Sea, with special emphasis on the Frisian Inlet, Ph.D. Thesis, Utrecht University.

Oost, A. P. and H. de Haas (1993). The Frisian Inlet, morphological and sedimentological changes in the period 1927–1970. Utrecht University, Institute of Earth Sciences, Coastal Genesis Report. 94 (in Dutch).

Peltier, E., J. Duplex, B. Latteux, P. Pechon and P. Chausson (1991). Finite element model for bed-load transport and morphological evolution. *Computer Modeling in Ocean Engineering 91*, eds. A. S.-Arcilla, M. Pastor, O. C. Zienkiewicz and B. A. Schrefler. Balkema. 227–233.

Pethick, J. S., 1980. Velocity surges and asymmetry in tidal channels. *Est. Coast. Mar. Sci.* **11**: 321–345.

Ribberink, J. S. and J. H. De Vroeg (1991). Coastal defence Texel (Eierland), hydraulic and morphological effect study; Phase 1: Morphological analysis. Delft Hydraulics, Report H 1241. Part I (in Dutch).

Ribberink, J. S. and J. H. De Vroeg (1992). Coastal Defence Texel (Eierland), hydraulic and morphological effects study; Phase 2/3: Morphological computations. Delft Hydraulics, Report H 1241. Parts III/IV (in Dutch).

Roelvink, J. A. and I. Brøker (1993). Cross-shore profile models. *Coast. Eng.* **21**(1-3): 163–191.

Roelvink, J. A. et al. (1994). Design and development of DELFT-3D and application to coastal morphodynamics. *Hydroinformatics '94*, eds. A. Verwey, A. W. Minns, V. Babović and M. Maksimović. Balkema. 451–456.

Schuttelaars, H. M. and H. E. De Swart (1994). A simple long-term morphodynamic model of a tidal inlet. Utrecht University, Department of Mathematics, Preprint 864.

Soulsby, R. L., L. Hamm, G. Klopman, D. Myrhaug, R. R. Simons and G. P. Thomas (1993). Wave-current interaction within and outside the bottom boundary layer. *Coast. Eng.* **21**(1-3): 41–69.

Speer, P. E. and D. G. Aubrey (1985). A study of non-linear tidal propagation in shallow inlet/estuarine systems: Part II: Theory. *Est. Coastal Shelf Sci.* **21**: 207–224.

Steijn, R. C. and G. Hartsuiker (1992). Morphodynamic reponse of a tidal inlet after a reduction in basin area. Delft Hydraulics, Coastal Genesis Report H 840.40.

Steijn, R. C., T. Louters, A. J. F. Van der Spek and H. J. de Vriend (1989). Numerical model hindcast of the ebb-tidal delta evolution in front of the Deltaworks. In *Hydraulic and Environmental Modeling of Coastal, Estuarine and River Waters*, eds. R. A. Falconer, et al. Gower Technical. 255–264.

Stelling, G. S. and J. J. Leendertse (1992). Approximation of convection processes by cyclic ADI-methods. *Proc. Conf. Estuarine and Coastal Modeling*, Tampa, Florida. 171–181.

Stive, M. J. F. (1986). A model for cross-shore sediment transport. *Coastal Engineering 1986 Proc.*, ed. B. L. Edge. ASCE. 1551–1564.

Teisson, C., M. Ockenden, P. Le Hir, C. Kranenburg and L. Hamm (1993). Cohesive sediment transport processes. *Coast. Eng.* **21**(1-3): 129–162.

Terwindt, J. H. J. and J. A. Battjes (1990). Research on large-scale coastal behavior. *Coastal Engineering 1990 Proc.*, ed. B. L. Edge. ASCE. 1975–1983.

Van de Kreeke, J. and K. Robaczewska (1993). Tide-induced transport of coarse sediment: Application to the Ems estuary. *Neth. J. Sea Res.* **31**(3): 209–220.

Van Dongeren, A. R. and H. J. de Vriend (1994). A model of morphological behavior of tidal basins. *Coast. Eng.* **22**(3&4): 287–310.

Van Rijn, L. C. (1984). Sediment transport; Part I: Bed load transport. *J. Hydr. Eng. ASCE* **110** (HY1): 1431–1456.

Van Rijn, L. C. (1989). Handbook sediment transport by currents and waves. Delft Hydraulics, Report H461.

Van Rijn, L. C. and K. Meijer (1991). Three-dimensional modeling of sand and mud transport in currents and waves. *The Transport of Suspended Sediment and its Mathematical Modeling, Int. IAHR/USF Symp.*, Florence, Italy. 683–708.

Wang, Z. B. (1992). Theoretical analysis on depth-integrated modeling of suspended sediment transport. *J. Hydr. Res.* **30**: 403–421.

Wang, Z. B., H. J. de Vriend and T. Louters (1991). A morphodynamic model for a tidal inlet. *Computer Modeling in Ocean Engineering 91*, eds. A. S.-Arcilla, M. Pastor, O. C. Zienkiewicz and B. A. Schrefler. Balkema. 235–245.

Wang, Z. B., T. Louters and H. J. de Vriend (1995). Morphodynamic modeling for a tidal inlet in the Wadden Sea. *Marine Geology.* **126**: 289–300.

Zitman, T. J. (1992). Quasi-three-dimensional current modeling based on a modified version of Davies' shapefunction approach. *Cont. Shelf Res.* **12**(1): 143–158.

BEACH-FILL DESIGN

JAMES R. HOUSTON

Beach nourishment involves placing sediment on the beach and in the nearshore area to mitigate coastal damage and beach erosion. A designed beach-nourishment project is called a beach fill. Beach fills can provide both storm-damage reduction and recreation benefits. They have been used throughout the world and are becoming increasingly popular because they provide ample benefits. This paper describes the information that must be collected to design a beach fill. Methods are provided to calculate required design variables. Two of the most widely used approaches to beach-fill design are presented along with many specific examples showing how to make necessary calculations. Beach fills need to be monitored after they are constructed, and the paper provides information on what should be monitored.

1. Introduction

1.1. *Mitigating coastal damage and beach loss*

There are three basic methods used to mitigate coastal damage or beach loss produced by storms or persistent long-term erosion. One can retreat as the shoreline retreats or as facilities are damaged during storms, "harden" the shoreline through the use of structures, or use sand to raise dunes and extend the shoreline seaward (known as beach nourishment). Often a combination of these methods is used at a particular location. The National Research Council (1990) notes that all of these methods can provide valid solutions depending on the specific causes of the erosion, the level of development, and other factors.

Retreat is the best method to mitigate damage and beach loss on a lightly populated coast, where the economic impact of damage or beach loss is relatively small and there is not a threat to a resource of significant economic importance (e.g., critical coastal road) or unique historic or ecologic value (e.g., historic coastal structure or habitat of an endangered species). Development can be moved away from the shoreline before it is endangered or, when damaged, facilities can be left in disrepair and eventually eliminated. A common form of retreat is to restrict new construction within a specified distance of the shoreline (sometimes called a "setback"). Retreat also is sometimes called the "do-nothing" option. It is the most commonly used option in the United States (US). In the US, about 33,000 kilometers of shoreline are eroding and about 4,300 kilometers are critically eroding. Beach protection and restoration projects involving hard structures or beach nourishment have been developed by the Corps of Engineers (the agency largely responsible for shore protection and restoration in the US) for only about 360 kilometers of shoreline (US Army Corps of Engineers, 1994). The reason the do-nothing option is the most widely used solution in the US is that for much of the shoreline the benefits of structures or nourishment are outweighed by their costs or funds are not available. However, for a developed shoreline, the retreat option can have the worst economic return. Although the retreat option is sometimes touted as being in tune with nature, there is nothing aesthetically pleasing about erosion damage eating into sewage systems or the beach littered with concrete from structural damage. Lack of a protective beach or structures in front of a developed shoreline can put development and even life at considerable risk.

Hard structures can be the best solution to coastal damage or erosion problems on a heavily populated coast, especially where the population or facilities are significantly threatened by storms. For example, about 6,000 people were killed by a hurricane at the turn of the century in Galveston, Texas, US. Galveston Island was raised and a major protective seawall built (Wiegel, 1991). The seawall has virtually eliminated loss of life in Galveston for almost a century including withstanding a hurricane in 1915 even larger than the hurricane of 1900 with little loss of life or damage to development. Certainly, retreat was a valid option for Galveston, and Galveston Island could have been left completely undeveloped. However, once development was allowed, construction of a protective structure was the only responsible solution to safeguard life. Given the significant economic activity at Galveston this century, the seawall has produced a substantial economic return.

On a developed coast, economics can strongly support a solution other than retreat. However, hard structures are often not considered desirable because of their possible impact on local or downdrift beaches. For example, structures such as groins that impound longshore transport of sand can be shown in many situations to cause downdrift erosion. Studies in recent years (Kraus, 1988; Griggs et al., 1994) have shown that seawalls that do not impact longshore transport do not actively produce erosion. However, as Tait and Griggs (1990) have discussed, using a structure to fix the location of a shoreline that is eroding or experiencing relative sea-level rise eventually leads to the shoreline reaching the structure. Although the structure may not be actively producing the erosion, the beach eventually disappears (sometimes called "passive" erosion) unless beach nourishment is used to counter losses. Seawalls that are well landward of the normal still-water level and have a protective function only during large storms have a minimal effect on beaches (Griggs et al., 1994).

Placing sediment to counter coastal damage or beach erosion is called beach nourishment. A designed beach-nourishment project is called a beach fill. Beach nourishment does not eliminate the problem causing shoreline recession. Erosion may result from reducing sediment availability by damming rivers supplying sediment, interrupting longshore transport through construction of updrift structures, opening of new inlets, deepening of navigation channels, relative sea-level rise, or other reasons. Nourishment restores the eroded beach, but erosion will still continue. The economic justification for nourishment is that restoring the beach results in economic gains as a result of increased recreational usage of the beach or increased protection of development. As long as benefits derived from restoring the beach exceed the cost of restoration, there can be a compelling economic justification for nourishment. Beach nourishment has become increasingly popular because it avoids possible problems of downdrift erosion that can be produced by structures. As a beach fill erodes, it provides material to downdrift beaches, providing additional benefits. This is the source of the Dutch observation that, "... one has to conclude that artificial beach nourishment has, from a financial point of view, a very big advantage. It is in fact an investment on which one has hardly any risk. Every grain of nourished sand is effective. Artificial beach nourishment may prove to be somewhat cheaper or more expensive than anticipated, but what has been invested is not lost". (Verhagen et al., 1994)

A key issue relating to beach nourishment is how well the performance of the fill can be predicted. If the fill erodes faster than projected, economic benefits from increased recreation will be reduced. However, if a fill protects a coastal location from a design storm that occurs sooner than its normal return period, the fill has provided an economic return from storm-damage reduction earlier than estimated. Although there can be some ecologic questions relating to beach nourishment (e.g., impact on sea turtle nesting or effects of turbidity during nourishment), the primary issue determining whether beach nourishment is the preferred option rather than retreat or structures is economic.

Therefore, proper beach-fill design is critical to maximize economic return and minimize risk.

Coastal damage and beach erosion problems usually are addressed in a region using a combination of the three basic methods. A coastal community may have a setback line to restrict future development. It may choose to do nothing along part of its coast because there is not a sufficient economic return in building structures or nourishing the beach along this part of the coast. Along the remainder of the coast, the best economic return may be obtained using structures or beach nourishment. Structures are often placed in conjunction with beach nourishment. For example, terminal groins are used to confine fills if ends of the fills are near sand sinks such as inlets or submarine canyons. The absence of a terminal structure can result in a large loss of sediment. Groins also may be placed in the interior of fills to slow the loss of sand. There are many examples of groins that have caused downdrift erosion. However, if properly designed with the overall sand budget of the region in mind, groins may provide an economic return with impacts minimized (National Research Council, 1990).

1.2. *Function of beach nourishment*

Beach nourishment provides protection from coastal storms and restores or enhances beaches. Beaches provide both storm-damage reduction and recreation benefits. The ability of increased beach width and dunes to reduce flooding and storm damage has been widely reported for both hurricanes (Shows, 1978; Stauble et al., 1991) and extratropical storms (Coastal Zone Management, 1992). Damage data following Hurricane Eloise show average damage costs suffered by structures increased rapidly with structure proximity to the shoreline (Shows, 1978). Beach nourishment can significantly reduce damage by increasing structure distance from the shoreline. For example, protection provided by a beach fill placed in 1986–1987 at Myrtle Beach, South Carolina, US, was evident in damage surveys following Hurricane Hugo in 1989. The surveys showed less damage in the fill-protected areas than in adjacent areas (Stauble et al., 1991). As a beach narrows, its recreation benefits drop. For example, before the major beach restoration project in the late 1970s at Miami Beach, Florida, US, the infrastructure at Miami Beach had deteriorated markedly because the narrow beach was not attractive to recreational-beach users. After the beach was nourished, the large increase in beach users increased the economic incentive to restore the infrastructure. The increase in recreation from the beach nourishment at Miami Beach is documented. Beach attendance, based on lifeguard counts and aerial surveys, increased from eight million in 1978 to 21 million in 1983 (Wiegel, 1992).

The economic importance of beaches is becoming more evident as the world economy moves increasingly toward service industries. Travel and tourism is the world's largest industry with revenues of US$2.9 trillion (Miller, 1993). The success of a country in competition for travel and tourism can have a significant impact on its international competitiveness. For example, the US had a US$17 billion (where a billion is a thousand million) trade surplus in travel and tourism in 1992 compared with a US$7-billion deficit in 1986 (*Wall Street Journal* 1993). Overseas tourists are expected to spend about US$80 billion in the US in 1994 (*Business Week* 1994). Miami Beach alone has two million overseas tourists a year who spend US$2 billion. This yearly expenditure compares with the cost of about US$50 million for the beach renourishment at Miami Beach. Beaches are the leading destination of tourists in the US with 40% of people listing beaches as a favorite vacation destination (*USA Today* 1993). Coastal states receive about 85% of all tourist-related revenues in the US (Houston, 1995). Stronge (1994) determined that just in the state of Florida, US, beach tourism creates US$15.4 billion in gross regional product, 359,000 jobs, and US$5-billion in wages.

1.3. *History of beach nourishment*

Beach nourishment has been used for shore protection and restoration for over 50 years. Almost 85 million cubic meters of sediment were placed on beaches in southern California, US, in about 60 projects between 1919 and 1978 (Herron, 1980). Only three of these projects were solely for beach nourishment. The primary purposes of most of these projects were development of harbors or excavation for coastal construction with the surplus sediment disposed on nearby beaches. This disposal transformed rocky and sand-starved beaches to wide beaches renowned throughout the world. Hall (1952) documents 72 beach nourishment projects in the US from 1922 to 1950. The US Army Corps of Engineers (1994) documents 56 major beach nourishment projects in the US for which about 130 million cubic meters of sediment were placed from 1959 to 1993.

Beach nourishment is a major tool for shore protection and restoration throughout the world. The Gold Coast, Queensland, Australia, has been nourished with about 5 million cubic meters of sediment (Smith and Jackson, 1990).

8.5 million cubic meters of sediment have been placed near Zeebruge, Belgium (Kerchaert et al., 1986). Between 1981 and 1987, over 9 million cubic meters of sediment was placed along 50 kilometers of coastline along the Black Sea (Kiknazde et al., 1990). Beach nourishment is the preferred approach for shore protection and restoration in Denmark with about 2.5 million cubic meters of sediment being placed yearly (Jakobsen and Sandgrav, 1994). Germany has spent about US$3.3 billion since 1955 for coastal protection of the German North Sea coast with beach nourishment the common approach for addressing coastal protection problems since 1972. The current expenditure rate for coastal protection along the German coast is US$100 million per year (Kelletat, 1992). The Netherlands placed about 60 million cubic meters of sediment in 50 projects from 1952 to 1989 (Roelvink, 1990). The Dutch adopted a coastline preservation policy in 1990 that calls for annual beach nourishment of about 6 million cubic meters per year at a cost of about US$32 million in 1990 price levels. The Dutch policy "... favors a periodic sacrificial nourishment because of its cost-efficiency, flexibility, and minimal environmental impact" (Verhagen et al., 1994). Dutch dredging contractors had over 200 beach nourishment projects worldwide between 1983 and 1992 and they estimated the market outside the US for nourishment at about US$100 million per year (Verhagen et al., 1994). The budget for shore protection and restoration in Japan was over US$1.5 billion in 1990 (Marine Facilities Panel, 1991). Spain placed over 50 million cubic meters of sand between 1983 and 1992 and plans restoration of its coast from 1993 to 1997 using 58 million cubic meters of sand and expenditure of over US$1 billion (Ministerio de Obras Publicas y Transportes, 1993).

2. Site Characterization

2.1. *Data requirements*

Planning, design, and construction of a beach-fill project requires collection of a variety of data. For example, it is important to know geomorphological characteristics of the area, sediment characteristics of the native beach and potential borrow areas, the extent and rate of erosion, the bathymetry of the site out to a water depth at which there can be significant movement of sediment during storms, waves at the site, and coastal processes including a sediment budget.

It is important to place a beach nourishment project in the context of the regional and local geomorphology that is affecting evolution of the coastline. For example, the Coastal Engineering Research Center (1992a) has developed guidance for using morphology to determine net littoral drift in complex areas. Regional geomorphological information can provide insight on large-scale changes affecting the beach-fill site, indicate sediment supply, and help determine possible borrow sites for beach-fill sediment.

Bathymetric measurements along lines perpendicular to the coast (called profile lines) are needed to determine the volume of material that will be required to advance the width of the beach to a desired amount. On beaches backed by dunes, the profile lines should extend across the primary dune and be far enough from the shoreline so that their baseline is not lost during future erosion, including erosion during episodic storms. The profiles should extend offshore to the depth of closure. The depth of closure is an approximate depth at which there is movement of sand of engineering significance. When making repetitive profile measurements over several years, it is the depth at which the change of elevation over repeated surveys falls within the accuracy of the measurement technique. The best method to measure bathymetry out to closure depth is with a device, such as a sea sled with a mast that can be used much like traditional surveying on land. The Coastal Engineering Research Center (1993a) provides guidelines for surveying beach nourishment projects.

Wave climatological data are important for fill design. For example, closure depth will be shown later to be related to the average mean wave height at the site. Wave climatology is important in determining the increase in the erosion rate over the native-beach erosion rate that results because a fill has a finite extent. It is important as input to numerical calculations to determine the effect of structures or interruptions in the sediment supply on the fill. Information on waves and water levels during episodic events is needed to help determine the width of fill needed to protect against a variety of storm events.

Historical-shoreline change is important to determine the life of a beach fill. After the fill has reached an equilibrium condition (discussed later), it will begin eroding at the historical rate plus an additional erosion (discussed later) that results from the finite length of the fill and the difference in sediment size between the borrow and native sediment. Historical maps and bathymetric charts can be used to estimate historical erosion rates. If profile measurements have been made for an extended period of time, this information can be used to determine erosion rates. Care should be taken to compare shorelines surveyed during the same time of the year to remove seasonal variations. The tidal stage when the measurements were made also must be known to accurately calculate the erosion rate.

A detailed study is required of sediment characteristics of the native beach and of borrow sites where fill sediment will be obtained. As will be shown later, the mean diameter of the fill sediment versus that of the natural beach is a critical design parameter. The suitability of borrow material for a beach fill is based on comparative analysis of the native-beach sediment and the borrow sediment. Sediment characteristics are usually collected along the profile lines discussed earlier. Grain size can be determined through sieve analysis or by observing settling velocities. Composite grain-size statistics can be determined using guidance in the Coastal Engineering Research Center (1991a). The Automated Coastal Engineering System (1992) provides computer software for composite grain-size analysis.

Locating economic borrow sources for a fill can be critical to the success of the fill. As will be shown later, the coarser the fill sediment, the less volume needed to establish a desired beach width. If the sediment is too fine, placement may result in no increase to the subaerial beach after the fill has reached an equilibrium condition. However, the distance the sediment must be transported and the method of transportation can greatly influence the cost of the fill; therefore, an analysis must be made of the economics of utilizing various borrow sources.

Borrow areas are typically terrestrial, back barrier, navigation channel, and offshore. Terrestrial sources are typically ancient fluvial and marine deposits. Truck transport is often used to move this sediment to the beach. The cost of truck transport is usually high and the noise and damage to roads is often undesirable. However, it is a flexible approach because mobilization costs are relatively low. Although costly for the initial nourishment, truck transport can sometimes be used effectively to maintain the beach fill through a continuous and relatively low level of truck transport. Sediment deposits in backbarrier marsh areas can be relatively cheaply exploited because the sediment is often close to the beach and a pipeline can be used. However, these sediments are often too fine grained for a beach fill. Sediment dredged from navigation channels is often sand of sufficient size to remain stable on a beach. Since the sediment must be moved out of the navigation channel and disposed of in any case, placement on a beach can sometimes be done at a relatively low price. Finally, offshore sediment is often used for beach fills. Ebb-tide shoals off inlets and relic linear and cape-associated shoals on the inner continental shelf often have sediment suitable for beach fills. The sediment can often be transported to shore relatively inexpensively using a dredge. It is important to ensure through calculations or hydraulic models that the borrow pit remaining after the sediment has been transported to shore does not modify the incident-wave field and produce erosion problems.

The locations of suitable borrow sediment are usually determined by first studying existing information on the geology of surrounding areas. Offshore areas are typically mapped using a boat with a fathometer and seismic-reflection surveys. Grab samples of surficial sediment and side-scan sonar might be used in the initial exploration phase. Sediment cores are then taken at the most promising locations. Exploration should determine limits of deposits, thickness of usable sediment, thickness of any overburden, and sediment characteristics.

2.2. Environmental considerations

Design of a beach fill should also consider environmental effects of the project. Rijkswaterstaat (1986) and Naqvi and Pullen (1982) summarize what is known about the effects of beach nourishment on the environment. Effects may occur at the fill or borrow site and can include turbidity generated at the fill or borrow site, direct impact on organisms through burial or the dredging operation, and impacts on nesting of sea turtles.

If the borrow sediment contains a significant quantity of fine-grain sediment, the fine material will be suspended in the water column at the borrow site and on the beach as it is reworked by waves. This turbidity may impact marine flora and fauna. Coral reefs and sea grasses can be especially harmed by fine sediments covering them. An analysis of ambient currents can be made to determine if the suspended fine sediments will drift to sensitive areas. It may be possible to choose a time of year when currents are such that the sediment does not drift to sensitive areas. If this cannot be done, it may be necessary to search for borrow material with less accompanying fine-grained sediment.

The beach-fill operation will directly affect organisms at the beach and borrow site. However, Stauble and Nelson (1985) show that the flora and fauna often recover relatively rapidly after a beach-fill operation. They provide guidelines for conducting operations to minimize impacts. Direct impact to mobile species such as fish are relatively small. Mortality of sessile flora and fauna that cannot vacate the area is relatively large, but recovery is usually rapid.

Sea turtles nest on beaches and their nesting can be impacted by beach-fill operations. Beach-fill operations are usually not conducted during nesting season on beaches that support turtle reproduction. Fill sediment may also be too compact and difficult for sea turtles to dig nests. Nelson *et al.* (1987) give methods to determine whether the sediment is too compact after placement. If it is, the beach can be tilled to loosen the sand to desired levels.

2.3. Equilibrium profile concept

The equilibrium profile of beaches is an important concept in beach-fill design. Bruun (1954) and Dean (1977, 1991) proposed that beach profiles develop a characteristic parabolic equilibrium beach shape given by

$$h = Ax^m, \qquad (1)$$

where h is the still-water depth, x is the horizontal distance from the shoreline, A is a dimensional parameter related to sediment characteristics, and $m = 2/3$. Bruun (1954) showed that Eq. (1) fits an extensive set of profiles from the Danish North Sea coast and a limited set of profiles from Mission Bay, California, US. Dean (1977) also found that Eq. (1) fits over 500 beach profiles collected by Hayden et al. (1975) along the US east coast and the Gulf of Mexico. Almost half of these profiles were from the east coast of Florida, US. Equation (1) has been compared favorably with measured data from Poland and the Black Sea (Pruszak, 1993) to New Zealand (Dean et al., 1993).

There have been numerous studies to develop conceptual models to explain the form of Eq. (1). An early paper by Keulegan and Krumbein (1949) concluded that a shoaling solitary wave would produce an equilibrium profile of the shape given by Eq. (1) with $m = 2/5$. Dean (1977) considered dissipation of shallow-water linear waves and showed that if equilibrium is associated with wave energy per unit area of bed, then $m = 2/5$. However, if equilibrium is associated with wave energy per unit volume of the water column, then $m = 2/3$. Dean (1977) concluded that $m = 2/3$ led to the best fit with field data of profiles.

Dean (1987) has shown that A can be related to the fall velocity of the sediment by the following:

$$A = 0.067 w^{0.44}. \qquad (2)$$

The fall velocity can be related to the sediment diameter using equations developed by Hallermeier (1981a). Hallermeier (1981a) gives fall-velocity equations for a wide range of beach sand, temperature, and both fresh and salt water. For the case of common beach sand with diameters in the range of 0.15 to 0.85 mm and temperatures from 15 to 25 degrees Celsius, Coastal Engineering

Research Center (1994a) shows that Hallermeier's equations can be reduced to give the following fall-velocity relationship:

$$w = 14\ D^{1.1}, \tag{3}$$

w has units of centimeters per second and D is the diameter of the median-grain size of sediment with units of millimeters (mm).

A can be expressed as

$$A = 0.21\ D^{0.48}. \tag{4}$$

For example, 0.2 mm sand has a fall velocity of 2.4 centimeters per second, which gives $A = 0.097$ meters to the one-third power.

3. Beach-Fill Design

3.1. Introduction

A beach fill usually has a design width that is the minimum needed to support the desired level of recreation or to provide a specified level of protection against storms. This width would be achieved after the first year's storms rework the fill sediment and the fill achieves an equilibrium condition. Fill design usually includes "advance nourishment" material placed to account for losses due to normal erosion at the site that creates the need for the fill, the finite length of the fill that increases the erosion rate over the background rate, and the difference in grain size between the borrow and native sediment. The interval between renourishment is dependent on the width of the advanced nourishment and the total erosion rate.

Early work on beach-fill design concentrated on the part of the beach that is visible, that is, the subaerial beach. However, it has been known for some time that the subaerial beach is only part of the beach system. When nourishing a beach, the entire nearshore profile down to an approximate closure depth must be nourished. Hallermeier (1981b) used laboratory data and limited field data from the Pacific Ocean and the Gulf of Mexico to develop a formula for computing closure depth. Birkemeier (1985) used an extensive data set measured at the Coastal Engineering Research Center's (CERC) Field Research Facility located along the Atlantic Ocean in northeastern North Carolina, US, to develop the following modification of Hallermeier's formula:

$$H = 1.75 H_{s0.137} - 57.9 \left(\frac{H_{s0.137}^2}{gT_s^2} \right), \tag{5}$$

where H is closure depth in meters, $H_{s0.137}$ is the extreme nearshore significant wave height exceeding 12 hours per year in meters (i.e., 0.137% of the year), and T_s is the period of the waves associated with $H_{s0.137}$. The Coastal Engineering Research Center (1994a) shows how Eq. (5) can be reduced to relate the closure depth simply to the mean annual significant wave height.

For a Pierson-Moskovitz spectrum (Rijkswaterstaat 1986),

$$\left(\frac{H_s^2}{gT_s^2}\right)^{1/2} = 6.4 \times 10^{-2}, \tag{6}$$

yielding

$$d = 1.5 \, H_{s0.137}. \tag{7}$$

From the Shore Protection Manual (1984),

$$F_{0.137} = \exp^{\frac{-(H_{s0.137} - H_{s\min})}{\sigma}}, \tag{8}$$

where

$F_{0.137}$ = the 0.137% exceedence frequency,

$H_{s\min} = H_s - \sigma$,

σ = the significant wave height standard deviation,

H_s = mean annual significant wave height,

$\exp$ = base of natural logarithms, 2.718.

Therefore,

$$H_{s0.137} = H_s + 5.6\sigma. \tag{9}$$

But $\sigma = 0.62 H_s$ from the Shore Protection Manual (1984) so

$$H = 1.5 \, H_{s0.137} = 6.75 H_s. \tag{10}$$

The relationship between the significant wave height standard deviation and mean annual significant wave height is approximate. When data are available, σ should be calculated directly rather than estimated using the relationship from the shore protection manual. The mean average significant wave height

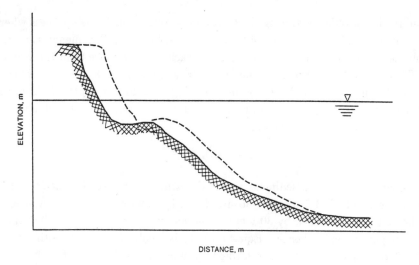

Fig. 1. Beach profile extended seaward uniformly. (Coastal Engineering Research Center, 1994a)

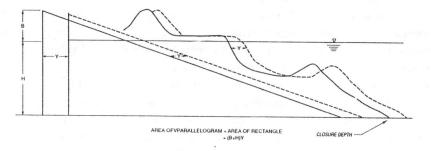

Fig. 2. Volume per unit length of shoreline from extending beach profile seaward. (Coastal Engineering Research Center, 1994a)

and σ for any location in the US is available from the a series of reports described by Coastal Engineering Research Center (1991b, 1991c, 1992b, 1993b).

If the size of beach-fill sediment is the same as that of the native beach, the beach profile after nourishment should be the same as that before nourishment except that it is extended seaward (Fig. 1) in a "reverse" Bruun rule. The Bruun rule was developed originally by Bruun (1962). It basically says that as sea level rises, the shoreline retreats uniformly so as to maintain a constant equilibrium profile. Restoring the beach is, therefore, an inverse process where the entire profile has to be built seaward. Of course, this means that the volume of sediment needed to advance a beach is much more than the volume of fill on the subaerial beach. The Shore Protection Manual (1984) shows that when the height of the beach berm is B (Fig. 2) and the depth of closure is H, to build out a beach, a distance, Y, requires a volume, V, of sediment per unit length of coastline given by

$$V = (B + H)Y. \qquad (11)$$

For example, if the fill-sediment size is the same as that of the native beach, the berm height is 1.5 meters, closure depth is 6 meters, and if one wishes to build out the dry beach by 30 meters, one has to place a volume of 225 cubic meters per meter of beach. Equation (11) is easily derived by noting from Fig. 2 that a parallelogram is formed by moving the beach out a distance, Y. The parallelogram can be shown to have an area identical to a rectangle (Fig. 2) with a base equal to $B+H$. Therefore, the volume of the fill per meter of coastline is equal to $(B+H)Y$.

The beach width of Y is achieved after equilibrium is reached usually following the first winter storm season. A common approach in the US, the Netherlands, and other countries is to initially place most fill volume on the subaerial portion of the active profile (Houston, 1991). Figure 3 shows a typical construction profile. Sediment is placed largely on the subaerial beach because this is usually the least expensive placement method. This placement method helps ensure that sediment is spread evenly along the beach. Nature is relied on to rework and distribute sediment along the complete active profile. Bruun (1988) has discussed cost advantages of placing part of the sediment along the offshore profile. Rijkswaterstaat (1986) discusses advantages in placing sediment mainly on the subaerial beach. Both Houston (1991) and Rijkswaterstaat (1986) point out the perception problem with initially placing most of the sediment on the subaerial beach as shown in Fig. 3. Even if the public is told what the final design beach width will be and further that the initial construction beach width is temporary, many will perceive a reduction of beach width (to the design width plus advance maintenance) over the first winter season to be a loss of beach. Houston (1991) shows that this effect has led to a perception among some in the US that beach fills have a poor success rate. On the other hand, local communities are often hesitant to pay for placement of sediment offshore rather than where they can see the payoff on the subaerial beach.

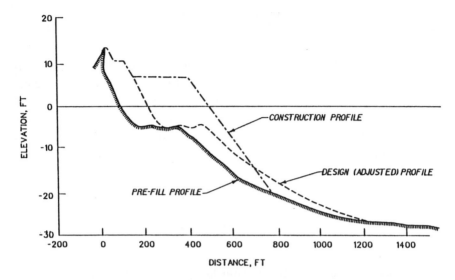

Fig. 3. Construction profile. (Coastal Engineering Research Center, 1994a).

Beach nourishment does not stop the pre-existing erosion problem. Therefore, periodic renourishment of the beach is necessary to maintain the designed beach width. The designed beach width, Y, is generally the minimum width desired at all times for recreation or storm-damage reduction. The minimum width of beach required for recreation is dependent on beach-usage levels. However, a minimum beach width of 30 meters is often considered ideal. In the US, most beach fills over the past decade or more have been designed to reduce storm damage because federal policies assign recreation a lower priority than flood control. Beach widths are determined based on the minimum width needed to reduce damages to a desired level.

Design of beaches to reduce storm damage generally requires a significant economic analysis. Since water levels are the dominant variable determining erosion during storms, increased water levels due to storms must be determined from historical data or by use of mathematical models using historic wind data to predict historic water levels (hindcasting). The probabilities of storm-induced increases in water level must be convolved with the fluctuating tide to produce combined probabilities of water-level stages. An engineering analysis must be coupled with an economic analysis to determine the economic damage expected for each water elevation. MacAllen and Cannon (1993) and Bodge (1991) give examples of analyses of beach-fill benefits relating to storm-damage reduction.

A key relationship needed to establish storm-damage reduction is the erosion expected for a given storm water level. If it is conservatively assumed that the water level is maintained long enough for equilibrium to be attained (rarely the case), the recession produced by a water level can easily be estimated. The net effect of increased water level is to move landward the origin of the equilibrium profile. From Dean (1983), the recession distance X for a storm-induced water level of S is given by

$$X = \frac{S(H/A)^{3/2}}{(H+B)}. \qquad (12)$$

For example, for a storm-induced water level $S = 2$ meters, $A = 0.1$ meters to the one-third power, $H = 6$ meters, and $B = 1.5$ meters, X is about 125 meters. This is an extremely large erosion, because it has been assumed that the profile has had sufficient time to reach equilibrium consistent with the peak-water level. However, the peak-water level exists for a relatively short time, and thus Eq. (12) greatly over predicts erosion. Mathematical models that take into account the time dependence of the storm elevation and other factors have been developed and should be used instead of Eq. (12) unless the large conservatism of Eq. (12) is required for situations where public safety is a major concern (such as for a levee protecting a major urban area).

Kriebel and Dean (1985) and Larson and Kraus (1989) present models that have been used extensively and consider the time factor in equilibrium-profile development. The SBEACH model described by Larson and Kraus (1989) is supported and available from the Coastal Engineering Research Center as is discussed by the Coastal Engineering Research Center (1994b).

Both erosion and damage increase with increased storm-induced water levels. Protection against increasing water levels requires increasing width of the beach and, therefore, increasing volume of fill. Storm-damage reduction benefits increase with increasing beach width, but costs of the fill also increase. Generally, the benefit-cost ratio increases with increasing beach width until a maximum is reached and then it declines. This maximum in the benefit-cost ratio then determines the fill width that provides the greatest economic return.

3.2. Dean's method for determining beach width for arbitrary sediment size

A key design parameter is the width of the dry beach after the beach profiles have reached equilibrium. For the case of fill sediment having a different size than the native sediment, Dean (1991) presents a method to determine the volume of sediment that must be placed per unit length of coast to achieve a desired dry-beach width after equilibrium. When fill sediment is identical to the native sediment and the distance out to closure depth is much larger than the width of the fill, Dean's method reduces to Eq. (11). The Coastal Engineering Research Center (1994a) provides guidance on applying Dean's method.

Dean (1991) defines three basic types of nourished profiles. Figure 4 shows an intersecting profile where the profile after nourishment intersects the native profile, a nonintersecting profile where the nourished profile does not intersect the native profile before closure depth, and a submerged profile where there is no dry beach after equilibrium. Dean (1991) shows that whether a profile is intersecting or nonintersecting is determined by the following inequalities:

$$Y \left(\frac{A_N}{H}\right)^{3/2} + \left(\frac{A_N}{A_F}\right)^{3/2} < 1, \text{ Intersecting profiles}, \qquad (13)$$

$$Y \left(\frac{A_N}{H}\right)^{3/2} + \left(\frac{A_N}{A_F}\right)^{3/2} > 1, \text{ Nonintersecting profiles}, \qquad (14)$$

where A_N equals the A value of the native sediment and A_F equals the A value of the fill sediment. For example, if $Y = 30$ meters, $H = 6$ meters, $A_N = 0.1$, and $A_F = 0.09$; then Inequality 14 is satisfied and the profiles are nonintersecting. However, if $A_N = 0.09$ and $A_F = 0.1$, then Inequality 13 is satisfied and the profiles are intersecting.

For the case of nonintersecting profiles where the fill sediment is finer than the native sediment (i.e., $A_F < A_N$), the volume of sediment that must be placed before there is any dry beach after equilibrium is

$$V = \frac{3}{5} H^{5/2} \left(\frac{1}{A_N}\right)^{3/2} \left(\frac{A_N}{A_F}\right)^{3/2} \left(\frac{A_N}{A_F} - 1\right), \qquad (15)$$

where V is the volume of sediment in cubic meters per meter of beach and H is the closure depth. If the volume placed is less than given by Eq. (15), the resulting profile is completely submerged.

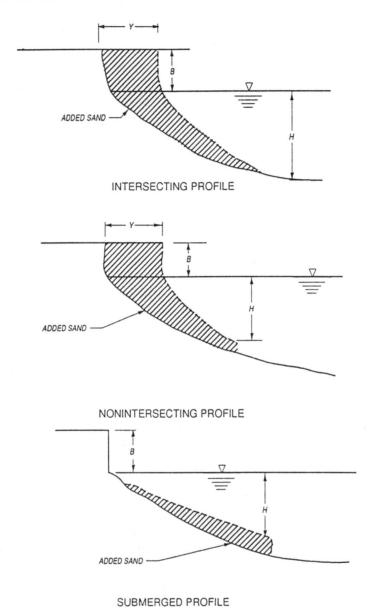

Fig. 4. Possible profiles using Dean's method. (Dean and Abramian, 1993)

For example, if $H = 6$ meters, $A_N = 0.1$, and $A_F = 0.09$, then $V =$ about 220 cubic meters per meter of beach. This means that this volume must be placed before there is any dry beach after equilibrium.

For nonintersecting profiles (satisfies Inequality 14) with a dry beach after equilibrium (i.e., volume placed is equal to or exceeds that in Eq. (15)),

$$V = YB + \frac{3}{5}H^{5/2}\left\{\left[\frac{Y}{H^{3/2}} + \left(\frac{1}{A_F}\right)^{3/2}\right]^{5/3} A_N - \left(\frac{1}{A_F}\right)^{3/2}\right\}. \quad (16)$$

For example, if $H = 6$ meters, $B = 1.5$, and the sediment diameters for the native beach and fill beaches are 0.25 mm and 0.2 mm respectively, what is the volume required to extend the width of the beach 30 meters after equilibrium? From Eq. (4), $A_N = 0.108$ and $A_F = 0.097$. Equation (16) yields a volume, V, of about 450 cubic meters per meter of beach length.

For intersecting profiles (generally the case when fill sand is coarser than the native, but Inequality 13 defines), the volume, V, required to advance the dry beach by a distance, Y, after equilibrium is given by

$$V = BY + \frac{A_N Y^{5/3}}{\left\{1 - \left(\frac{A_N}{A_F}\right)^{3/2}\right\}^{2/3}}. \quad (17)$$

For example, if the berm height is 1.5 meters, the native and fill beaches have sediment sizes of 0.2 and 0.25 mm respectively (so the fill sand is coarser than the native), and we wish to extend the dry beach 30 meters after equilibrium; $A_N = 0.097$ and $A_F = 0.108$. From Equation (17), V equals about 145 cubic meters per meter of beach. Note that the closure depth is not contained in Eq. (17) since the profiles intersect before the closure depth. The problem is essentially the same as that in the previous paragraph except that the grain sizes of the native and fill sands are reversed. The fill with coarser sediment than the native requires a substantially smaller volume than the fill with finer sediment than the native, as expected. Recall that when the native and fill sediments are the same, the required fill volume of 225 cubic meters per meter determined using Eq. (11) is a value that lies between the volumes determined using Eq. (16) and Eq. (17).

A shortcoming of the approach presented is that it represents the sediment on the profile by a single median-grain size. Actually, it is well known that sediment size generally decreases as one moves seaward. Dean and Abramian (1993) modifies the approach given above to allow a decrease in sediment size along the profile.

In summary, when using Dean's method, the berm height, B, and width of the dry beach after equilibrium, Y, are given. The closure depth, H, equals 6.75 times the mean average significant wave height. Sediment sizes for the native and fill beaches are known. A_N and A_F are determined using Eq. (4). Inequalities 13 and 14 are used to determine if the profiles are intersecting or nonintersecting. If they are intersecting, Eq. (17) is used to calculate the volume of sediment required to advance the shoreline by a distance Y. If the profiles are nonintersecting, Eq. (16) is used to calculate Y. Equation (15) can be used for nonintersecting profiles to determine the volume that must be placed to make Y greater than zero. Examples are provided for each case in the paragraphs above.

3.3. *Overfill ratio method*

When beach-fill material is different than the native material, Krumbein (1957), Krumbein and James (1965), James (1974), Dean (1974), and James (1975) have developed similar approaches to indicate probable behavior of fill material. These approaches develop an overfill ratio that indicates how much extra sediment is required to be placed as a result of the borrow material having different sediment characteristics than the native material. The overfill-ratio approach assumes that both the native and borrow grain have a lognormal distribution and the native sediment on the beach is the most stable for its environment. Further, the fill sediment is sorted by wave action until it adopts a grain-size distribution similar to the native sediment. Sediment finer than the native sediment will be lost. With these assumptions, James (1975) provides a method for estimating the required additional volume of sediment when the borrow sediment differs from the native sediment. The Shore Protection Manual (1984) gives a detailed description of the method. Although the overfill ratio has been used extensively in beach-fill design, the Shore Protection Manual (1984) notes that, "It should be stressed, however, that these techniques have not been fully tested in the field and should only be used as a general indication of possible beach-fill behavior."

Important parameters used in this approach are values for the composite mean and standard deviation of sediment diameter. The overfill ratio is determined by comparing ϕ mean and sorting values of the native and borrow beach sediments where ϕ units are equal to minus the base two logarithmic value of the sediment diameter in millimeters (mm).

$$\phi = -\log_2(D) = \frac{-\ln(D)}{\ln 2} = \frac{-\log_{10}(D)}{\log_{10} 2} \qquad (18)$$

where ln is the natural logarithm. For example, a 0.2 mm sand has a ϕ value of 2.3.

The overfill ratio is the volume of borrow sediment required to produce a single unit of usable fill sediment with the same grain size characteristics as the native sediment. The equation for determining the overfill ratio is complex and a graphical method is usually used (Fig. 5). In Fig. 5, the following define terms:

b = subscript refers to borrow sediment,

n = subscript refers to native sediment on the beach,

$M_\phi = (\phi_{84} + \phi_{16})/2$, is the ϕ mean diameter of grain-size distribution,

$\sigma_\phi = (\phi_{84} - \phi_{16})/2$, is the ϕ standard deviation and is a measure of sorting,

ϕ_{84} = 84th percentile in ϕ units,

ϕ_{16} = 16th percentile in ϕ units.

As an example, suppose the native and borrow beach sediment have the following characteristics:
Native Beach Sediment

$\phi_{84} = 2.47\phi(0.18 \text{ mm})$,

$\phi_{16} = 1.41\phi(0.38 \text{ mm})$.

Fill Beach Sediment

$\phi_{84} = 3.41\phi(0.09 \text{ mm})$,

$\phi_{16} = 1.67\phi(0.31 \text{ mm})$.

Therefore,

$M_{\phi n} = (2.47 + 1.41)/2 = 1.94$,

$M_{\phi b} = (3.41 + 1.67)/2 = 2.54$,

$\sigma_{\phi n} = (2.47 - 1.41)/2 = 0.53$,

$\sigma_{\phi b} = (3.41 - 1.67)/2 = 0.87$.

Using Fig. 5 the overfill ratio R equals about 2.5.

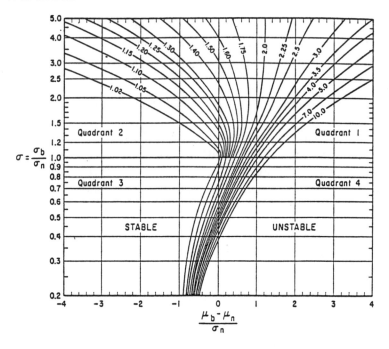

Fig. 5. Isolines of the adjusted fill factor versus phi mean difference and phi shorting ratio. (James, 1975)

The adjusted overfill ratio can be determined directly using the following equation:

$$\frac{1}{R} = 1 - erf\left(\frac{\theta_2 - \delta}{\sigma}\right) + erf\left(\frac{\theta_1 - \delta}{\sigma}\right) + \left[\frac{erf(\theta_2) - erf(\theta_1)}{\sigma}\right] \cdot$$
$$\exp\left\{\frac{1}{2}\left[\theta_1^2 - \left(\frac{\theta_1 - \delta}{\sigma}\right)^2\right]\right\} . \quad (19)$$

Where

$$\delta = \frac{M_{\theta b} - M_{\theta n}}{\sigma_{\theta n}}, \quad (20)$$

$$\sigma = \frac{\sigma_{\theta b}}{\sigma_{\theta n}} . \quad (21)$$

When $\sigma > 1 : \theta_1 = $ maximum of $\{-1$ or $\frac{-\delta}{\sigma^2 - 1}\}; \theta_2 = \infty$.

When $\sigma < 1 : \theta_1 = -1; \theta_2 =$ maximum of $\{-1$ or $1 + \frac{2\delta}{1-\sigma^2}\}$, erf is the error function and when $x > 0, erf(x) = \frac{2}{\sqrt{\pi}} \int_0^x e^{-\frac{t^2}{2}} dt$.
When $x < 0, erf(x)$ is replaced by $1 - erf(x)$. When $x \to \infty, erf(x) \to 1$.

The erf function can be evaluated by using tables in Abramowitz and Stegun (1972) or the following equations can be used (Bajpai et al., 1977).

When $x \ll 1$, a useful formula is

$$erf(x) = \frac{2}{\sqrt{\pi}} \left(x - \frac{x^3}{3} + \frac{x^5}{5 \cdot 2!} - \frac{x^7}{7 \cdot 3!} + \cdots \right). \quad (22)$$

Where ! is the factorial sign so that $3! = 3 \cdot 2 \cdot 1$.

When $x \gg 1$, a useful formula is

$$erf(x) = 1 - \frac{e^{-x^2}}{x\sqrt{\pi}} \left(1 - \frac{1}{2x^2} + \frac{1 \cdot 3}{(2x^2)^2} - \frac{1 \cdot 3 \cdot 5}{(2x^2)^3} + \cdots \right). \quad (23)$$

These formulas hold for all values of x, but converge rapidly only for the limits given period. Equation (22) actually converges fairly rapidly when x is approximately one or less.

In the above example, $\sigma = 1.64$ and $\delta = 1.13$. Therefore, $\theta_1 =$ maximum of $\{-1$ or $-0.669\} = -0.669$.

$$\theta_2 = \infty \text{ and } erf\left(\frac{\theta_2 - \delta}{\sigma}\right) = erf(\infty) = 1.$$

So,

$$\frac{1}{R} = 1 - 1 + erf(-1.097) + \left[\frac{1 - erf(-.669)}{1.64}\right] \cdot$$

$$\exp\frac{1}{2}\left[(-.669)^2 - \left(\frac{-.669 - 1.13}{1.64}\right)^2\right]$$

$$\frac{1}{R} = erf(-1.097) + \left[\frac{1 - erf(-.669)}{1.64}\right](.685)$$

$$\frac{1}{R} = 1 - erf(1.097) + \frac{erf(.669)}{1.64}(.685).$$

But $erf(1.097) \approx 0.88$ and $erf(0.669) \approx 0.65$. Therefore, $R \approx 2.5$. This is the same value obtained earlier using Fig. 5.

The overfill ratio is generally combined with Eq. (11) to determine the volume of fill required to advance the dry beach by a specific distance. Therefore, the volume required to build the beach out a distance Y is given by

$$V = (B + H)RY. \tag{24}$$

For example, if the beach berm height is 1.5 meters, closure depth is 6 meters, and the beach is to be built out 30 meters, Eq. (21) gives the volume of sediment required as 225 cubic meters per meter of beach. With the native and fill sediments having the characteristics given above, the overfill ratio indicates that about 560 cubic meters of sediment needs to be placed per meter of beach.

3.4. Advance maintenance

As discussed earlier, additional material is usually placed in a fill so that when it has eroded, the beach is at the minimum design width and is ready to be renourished. The volume required for advance nourishment depends on the background erosion rate before the fill, size of the native and borrow sediments, the length of the fill, and the time desired between renourishments. Since mobilization of equipment may be a significant portion of renourishment costs, the renourishment interval is usually relatively long. Typical renourishment intervals are 4 to 10 years.

Advance maintenance must account for the background erosion rate that was occurring before nourishment. Generally, this rate is obtained by studies of historical data. If the sediment of the fill is identical to the native sediment, background erosion expected between nourishments is obtained by multiplying the erosion rate by the renourishment interval. The volume of sediment per meter of beach that must be placed to account for the background rate is obtained from Eq. (11). The erosion rate is multiplied by the renourishment interval to give a width of beach in meters and this in turn is multiplied by $(B + H)$. When the overfill ratio approach is used, this width is further multiplied by the overfill ratio.

To this point the assumption has been that the fill has an infinite length. Fill length has a significant effect on fill longevity. Pelnard-Considere (1956) combined a linearized equation of sediment transport and a continuity equation considering profiles to be displaced without change of form to yield a longshore-diffusion equation. Dean and Yoo (1993) further developed this approach to determine fill loss due to the finite length of the fill. This loss is often the dominant fill loss. Consequently, the best method to include the loss due to a finite fill length is to use a shoreline-response numerical model such as a one-line model. Several one-line models have been developed. The Coastal Engineering Research Center supports the publicly-available model GENEralized Model for SImulating Shoreline Change (or GENESIS). Shoreline-response models also can calculate the effect of structures on longevity of the fill. For the case of a beach fill placed on an infinite straight beach with no structures, Dean and Yoo (1993) combine an equation for sand conservation with a sediment-transport equation and develop a linearized equation for beach planform evolution. This equation can be used to estimate the effects of a finite length fill on fill life. Expressed in terms of the fraction of sediment remaining in the location placed as a function of time, $M(t)$,

$$M(t) = \frac{1}{J\sqrt{\pi}}(e^{-J^2} - 1) + erf(J), \tag{25}$$

where

$$J = \frac{L_f}{2\sqrt{Gt}}, \tag{26}$$

$erf(J)$ is the error function, t is time (in seconds), and L_f is the length of the fill.

$$G = \frac{(0.1 h_b^{5/2} g^{1/2})}{(H + B)}, \tag{27}$$

When $\frac{1}{J}$ is less than 1.0 ($M(t)$ being between 1.0 and 0.5),

$$M(t) = 1 - \frac{1}{J\sqrt{\pi}}, \tag{28}$$

where this is an approximate relationship.

The error function, $erf(J)$, can be found in mathematical tables such as Abramowitz and Stegun (1972) or calculated using Eqs. (22) or (23).

Losses due to the finite length of a fill can be mitigated by extending the lateral ends of a project at some angle to taper smoothly with the prefill shoreline. The Coastal Engineering Research Center (1985) presents an analytical solution to aid in tapering the fills to minimize losses. Hanson and Kraus (1993) show that the analytical solution overestimates the optimum length of the taper section, and that a shoreline response model such as GENESIS is the preferred method to make calculations on tapering fills.

3.5. Monitoring

Monitoring a beach fill after it is constructed is a central feature of beach-fill design. Since the fill does not eliminate factors causing beach erosion, eventually the beach will have to be renourished. Just as a road or any other infrastructure requires maintenance, a beach that is nourished will require future maintenance through renourishments. It is critical, therefore, that fills be monitored to learn if they are performing as expected or, if not, what can be done to improve the performance of future nourishments. Monitoring also is used to adjust nourishment schedules. Monitoring will identify whether the fill is eroding at the expected rate or whether the nourishment schedule must be adjusted to accommodate greater or lesser erosion. It also will identify "hot spots." Hot spots are areas along the shoreline of the fill experiencing greater erosion than adjacent areas. The reason for the greater erosion may not be known. However, erosional hot spots may require the beach or portions of it to be nourished earlier than predicted, and future nourishments may place extra sediment at these locations.

Beach fills have often not been monitored or inadequately monitored including monitoring for too short a period. Consequently, there has been considerable debate on how well many fills have functioned. The need for monitoring is now universally recognized and recommended. A monitoring plan should be developed before the fill is built and funding set aside to monitor the fill over an extended period.

The Coastal Engineering Research Center (1991d) provides guidance for physical data collection of beach nourishment projects including pre- and post-fill monitoring programs. The minimum monitoring program should include monitoring of the width of the beach over time. This monitoring allows one to determine the erosion rate of the beach for evaluation of the original design and to determine when renourishment is necessary. In addition, hot spots of erosion along the fill can be identified and mitigated if needed.

Simple monitoring of the width of the beach is the minimum monitoring that should be conducted, but it does not provide information on the fate of much of the beach fill nor the processes affecting the fill. Much of the sediment of a fill adjusts so that it is on the subaqueous portion of the beach. Determining the fate of this sediment requires hydrographic surveys. The

Coastal Engineering Research Center (1993c) provides guidelines for surveying beach nourishment projects. A direct measurement of depth such as using a sea sled with a mast (making the accuracy of this surveying technique similar to land surveys) is the best method to use to monitor beach fills. The Coastal Engineering Research Center (1991d) and Stauble and Nelson (1985) provide additional recommendations relating to sediment sampling and analysis, wave measurements, and biological monitoring.

4. Acknowledgements

The author wishes to acknowledge the Office, Chief of Engineers, US Army Corps of Engineers, for authorizing publication of this chapter.

References

Abramowitz, M. and I. A. Stegun (1972). *Handbook of Mathematical Functions with Formulas, Graphs, and Mathematical Tables*, Applied Mathematical Series 55. National Bureau of Standards, USA.

Automated Coastal Engineering System (1992). *User's Guide and Technical Reference*. Coastal Engineering Research Center, US Army Engineer Waterways Experiment Station, Vicksburg, Mississippi, USA.

Bajpai, A. C., Mustoe, L. R., and D. Walker (1977). *Advanced Engineering Mathematics*. John Wiley and Sons. 578.

Birkemeier, W. A. (1985). Field data on seaward limit of profile change. *J. Wtrwy., Port, Coast. and Oc. Eng. ASCE.* **111**(3):598–602.

Bodge, K. R. (1991). Damage benefits and cost sharing for shore protection projects. *Shore and Beach* (April):11–18.

Bruun, P. (1954). Coastal erosion and development of beach profiles, US Army Beach Erosion Board, US Army of Engineers, Washington, D.C., USA, Technical Memorandum No. 44.

Bruun, P. (1962). Sea Level Rise as a Cause of Shore Erosion. *J. Wtrwys. Harbors Div.* **1**:116–130.

Bruun, P. (1988). Profile renourishment: It's background and economic advantages. *J. Coast. Res.* **4**:219–228.

Business Week (1994). Oh say can we spend, a weak dollar is luring tourists to the US in droves, 22 August:30–31.

Coastal Engineering Research Center (1991a). Native beach assessment techniques for beach fill design, US Army Engineer Waterways Experiment Station, Vicksburg, Mississippi, USA, Technical Note II-29.

Coastal Engineering Research Center (1991b). Evaluation and application of the wave information study for the Gulf of Mexico. US Army Engineer Waterways Experiment

Station, Vicksburg, Mississippi, USA, Technical Note I–48.

Coastal Engineering Research Center (1991c). Evaluation and application of the wave information study for the Northern Pacific Ocean. US Army Engineer Waterways Experiment Station, Vicksburg, Mississippi, USA, Technical Note I–49.

Coastal Engineering Research Center (1991d). Recommended physical data collection program for beach nourishment projects. US Army Engineer Waterways Experiment Station, Vicksburg, Mississippi, USA, Technical Note II–26.

Coastal Engineering Research Center (1992a). Using morphology to determine net littoral drift directions in complex coastal systems. Coastal Engineering Research Center, US Army Engineer Waterways Experiment Station, Vicksburg, Mississippi, USA, Technical Note II–29.

Coastal Engineering Research Center (1992b). Revised wave information study (WIS) results for the US Atlantic Coast 1956–1975. US Army Engineer Waterways Experiment Station, Vicksburg, Mississippi, USA, Technical Note I–51.

Coastal Engineering Research Center (1993a). Guidelines for surveying beach nourishment projects. Coastal Engineering Research Center, US Army Engineer Waterways Experiment Station, Vicksburg, Mississippi, USA, Technical Note II–31.

Coastal Engineering Research Center (1993b). Availability of NDBC/NOAA data at WES. US Army Engineer Waterways Experiment Station, Vicksburg, Mississippi, USA, Technical Note I–55.

Coastal Engineering Research Center (1993c). Guidelines for surveying beach nourishment projects. US Army Engineer Waterways Experiment Station, Vicksburg, Mississippi, USA, Technical Note II–31.

Coastal Engineering Research Center (1994a). Beach-fill volume required to produce specified dry beach width. Coastal Engineering Research Center, US Army Engineer Waterways Experiment Station, Vicksburg, Mississippi, USA, Technical Note II–32.

Coastal Engineering Research Center (1994b). Application of SBEACH to coastal projects. Coastal Engineering Research Center, US Army Engineer Waterways Experiment Station, Vicksburg, Mississippi, USA, Technical Note II–33.

Coastal Zone Management (1992). Beach sand replenishment has received a hearty endorsement from Ocean City, Maryland **23**(12):1.

Dean, R. G. (1974). Compatibility of borrow material for beach fills. *Proc. 14th International Conf. Coastal Eng.* ASCE. 1319–33.

Dean, R. G. (1977). Equilibrium beach profiles: US Atlantic and Gulf Coasts. Department of Civil Engineering, University of Delaware, Newark, Delaware, USA, Ocean Engineering Report No. 12.

Dean, R. G. (1983). Shoreline erosion due to extreme storms and sea level rise. Department of Coastal and Oceanographic Engineering, University of Florida, Gainesville, Florida, USA, Report UFL/COEL-83/007.

Dean, R. G. (1987). Coastal sediment processes: Toward engineering solutions. *Proc. Specialty Conf. Coastal Sediments '87*. ASCE. 1–24.

Dean, R. G. (1991). Equilibrium beach profiles: Characteristics and applications. *J. Coast. Res.* **7**(1):53–84.

Dean, R. G. and C. Yoo (1993). Predictability of beach nourishment performance. *Beach Nourishment and Management Considerations.* ASCE. 86–192,.

Dean, R. G., Healy, T. R., and A. P. Dommerholt (1993). A 'blind-folded' test of equilibrium beach profile concepts with New Zealand data. *Marine Geology* **109**:253–266.

Dean, R. G. and J. Abramian (1993). Rational techniques for evaluating the potential of sands for beach nourishment. Coastal Engineering Research Center, US Army Engineer Waterways Experiment Station, Vicksburg, Mississippi, USA, Technical Report DRP-93-2.

Griggs, G. B., Tait, J. F., and W. Corona (1994). The interaction of seawalls and beaches: Seven years of monitoring, Monterey Bay, California. *Shore and Beach* **62**(3): 21–28.

Hall, J. V. (1952). Artificially constructed and nourished beaches in coastal engineering. *Proc. 3rd Coastal Eng. Conf. ASCE.*

Hallermeier, R. J. (1981a). Fall velocity of beach sands. Coastal Engineering Research Center, US Army Engineer Waterways Experiment Station, Vicksburg, Mississippi, USA, Technical Note CETN-II-4.

Hallermeier, R. J. (1981b). Seaward limit of significant sand transport by waves: An annual zonation for seasonal profiles. Coastal Engineering Research Center, US Army Eng. Waterways Experiment Station, Vicksburg, Mississippi, USA, Technical Aid No. 81–2.

Hanson, H. and N. C. Kraus (1993). Optimization of beach fill transitions. *Beach Nourishment Eng. and Management Considerations.* ASCE. 103–117.

Hayden, B., W. Felder, J. Fisher, D. Resio, L. Vincent, and R. Dolan (1975). Systematic variations in inshore bathymetry. Department of Environmental Science, University of Virginia, Charlottesville, Virginia, USA, Technical Report 10.

Herron, W. J. (1980). Artificial beaches in Southern California. *Shore and Beach* **48**(1):3–12.

Houston, J. (1991). Beachfill performance. *Shore and Beach* **59** (3):15–24.

Houston, J. (1995). Beach nourishment. *Shore and Beach* **63** (1):21–24.

Inman, D. L., H. S. Elwany, and S. A. Jenkins (1993). Shorerise and bar-berm profiles on ocean beaches. *J. Geophysical Res.* **98**(c):18181–18199.

Jakobsen, P. R. and B. Sandgrav (1994). Modern beach nourishment in Denmark — How to do it. *Proc. Permanent International Association of Navigation Congresses*, Section II, Subject 4, Seville, Spain, 17–22.

James, W. R. (1974). Borrow material texture and beach fill stability. *Proc. 14th International Conf. Coastal Eng. II.* ASCE. 1334–1344.

James, W. R. (1975). Techniques in evaluating suitability of borrow material for beach nourishment. Coastal Engineering Research Center, US Army Engineer Waterways

Experiment Station, Vicksburg, Mississippi, USA, Technical Report No. 60.

Kelletat, D. (1992). Coastal erosion and protection measures at the German North Sea Coast. *J. Coastal Res.* **8**(3):699–711.

Kerchaert, P., P. P. L. Roobers, A. Noordam, and P. DeCandat (1986). Artificial beach nourishment on Belgium east coast. *J. Wtrwy, Port, Coast. and Oc. Eng. ASCE.* **112**(5):560–671.

Keulegan, G. H. and W. C. Krumbein (1949). Stable configuration of beach slope in a shallow sea and its bearing on geological processes. *EOS Trans.* **30**(6):855–861.

Kiknazde, A. G., V. V. Sakvarelidze, V. M. Peshkov, and G. E. Russo (1990). Beach-forming process management of the Georgian Black Sea Coast. *J. Coast. Res.*, **6**(6):33–44.

Kraus, N. C. (1988). The effects of seawalls on the beach: An extended literature review. *J. Coastal Res.* **6**(4).

Kriebel, D. L. and R. G. Dean (1985). Numerical simulation of time-dependent beach and dune erosion. *Coast. Eng.* **9**(3):221–245.

Krumbein, W. C. (1957). A method for specification of sand for beach fills. US Army Corps of Engineers, Beach Erosion Board, Washington DC, USA, Technical Report TM–102.

Krumbein, W. C. and W. R. James (1965). Spacial and temporal variations in geometric and material properties of a natural beach. Coastal Engineering Research Center, US Army Engineer Waterways Experiment Station, Vicksburg, Mississippi, USA, Technical Report No. 44.

Larson, M. and N. C. Kraus (1989). SBEACH: Numerical model for simulating storm-induced beach change, Report 1, Empirical Foundation and Model Development. Coastal Engineering Research Center, US Army Engineer Waterways Experiment Station, Vicksburg, Mississippi, USA, Technical Report CERC-89-9.

MacAllen, T. C. and M. G. Cannon (1993). Modeling beach erosion control benefits. *The Military Engineer* no. 560, November–December, 1993, 664–67.

Marine Facilities Panel (1991). US/Japan cooperative program in natural resources, *Proc. 17th Joint Meeting*, **J**–(4):1–9.

Miller, M. L. (1993). The rise of coastal and marine tourism. *Oc. and Coast. Managemt.* **20**:181–199.

Ministerio de Obras Publicas y Transportes (1993). *Recuperando La Costa, Serie Monografias*. Centro de Publicaciones.

Naqvi, S. M. and E. J. Pullen (1982). Effects of beach nourishment and borrowing on marine organisms. Coastal Engineering Research Center, US Army Engineer Waterways Experiment Station, Vicksburg, Mississippi, USA, Miscellaneous Report. 82–14.

National Research Council (1990). *Managing Coastal Erosion*. National Academy Press.

Nelson, D. A., K. Mauck, and J. Fletemeyer (1987). Physical effects of beach nourishment on sea turtles, Delray, Beach, Florida. US Army Engineer Waterways Experiment Station, Vicksburg, Mississippi, USA, Technical Report EL-87-15.

Pelnard-Considere, R. (1956). Essai de Theorie de l'Evolution des Formes de Rivate en Plages de Sable et de Galets. *4th Journees de l'Hydraulique*, Les Energies de la Mar, France, Question III, Rapport No. 1.

Pruszak, Z. (1993). The analysis of beach profile changes using Dean's method and empirical orthogonal functions. *Coast. Eng.* **19**:245–261.

Rijkswaterstaat (1986). *Manual on Artificial Beach Nourishment*. Delft Hydraulics Laboratory, Netherlands.

Roelvink, P. (1990). Beach and dune nourishment in the Netherlands. *Proc. 22nd International Coastal Eng. Conf. ASCE.* 1984–1997.

Coastal Engineering Research Center (1984). *Shore Protection Manual*. Coastal Engineering Research Center, US Government Printing Office, Washington DC, USA. 4th ed.

Shows, E. W. (1978). Florida's coastal setback line — An effort to regulate beachfront development. *J. Coastal Zone Management* **4**(1/2):151–164.

Smith, A. W. and L. A. Jackson (1990). An application of coastal management tactics, Gold Coast, Queensland, Australia. *Shore and Beach* **58**:3–8.

Stauble, D. K., W. C. Seabergh, and L. Z. Hales (1991). Effects of Hurricane Hugo on the South Carolina Shore. *J. Coast. Res.* **8**:129–162.

Stauble, D. K. and W. G. Nelson (1985). Guidelines for beach nourishment: A necessity for project management. ASCE. Coastal Zone 85, Baltimore, MD, USA.

Stronge, W. B. (1994). Beach, tourism and economic development. *Shore and Beach* **62** (2):6–8.

Tait, J. F. and G. B. Griggs (1990). Beach response in the presence of a seawall — A comparison of field observations. *Shore and Beach* **58**(2):11–28.

USA Today (1993). *More Plan Vacations This Year*. 17 March, D1.

US Army Corps of Engineers (1994). Shoreline protection and beach erosion control study, Phase I: Cost comparison of shoreline protection projects of the US Army Corps of Engineers. Water Resources Support Center, Washington DC, USA, IWR Report 94-PS-1 January 1994.

Verhagen, H. J., K. W. Pilarczyk, and G. G. A. Loman (1994). Dutch approach to coastline management. *Proc. Permanent International Association of Navigation Congresses*, Seville, Spain, Subject II, Subject 4, 69–90.

Wall Street Journal (1993). *US Exports are Growing Rapidly But Almost Unnoticed*. 21 April, A1.

Wiegel, R. L. (1991). Protection of Galveston, Texas, From overflows by gulf storms: Grade-raising, seawall and embankment. *Shore and Beach* **59**(1).

Wiegel, R. L. (1992). Dade County, Florida, beach nourishment and hurricane surge study. *Shore and Beach* **60**(4):2–26.

SHEAR CURRENT AND ITS EFFECT ON FIXED AND FLOATING STRUCTURES

SUBRATA K. CHAKRABARTI

Structures are often installed in environments which include a high velocity current field. Examples of these structures in confined flowing waters are underwater pipelines, gravity structures, floating barges, and tankers. The effect of the current flow on these structures is important in designing the foundations and/or mooring systems for these structures. The current may be uniform or varying with depth. The varying currents are of two types. Some show higher speeds at the surface and decrease with depth, while others have the reverse trend. It is often found in nature that these currents can be approximated by a linear shear. The shear strength in these cases are defined by the slope of the current speed with depth. It has been found that the loads on a structure in shear current are quite different from those in uniform flow. This chapter summarizes the effects of the uniform and shear current flow on fixed and floating structures and provides information on their loadings.

1. Introduction

When blunt bodies are placed within a flowing fluid (liquid or gaseous) medium, the fluid flow pattern is altered and restraints are required to maintain the position of the body within the fluid medium. The loads on the body are normally characterized as drag (in-line with the direction of flow) and lift (transverse to the direction of flow). In either direction, the load can often be described as being composed of a relatively steady component and higher frequency components which are functions of the body shape and flow conditions.

Downstream of the blunt object, the fluid flow regime will eventually return to its upstream unaltered condition due to fluid viscosity and damping. The region of altered flow directly behind the body is defined as the "wake", and many investigators have demonstrated the unique relationship that exists between the wake and the restraint loads. The fluid-structure interaction has been shown to be determinant, i.e., for the same body and flow conditions, the restraint loads will have the same steady and higher frequency components. The wake, although seemingly chaotic, will have a similar width and will have the same frequency content as the restraint loads. In this regard, the body is defined by its shape, surface roughness and orientation. The fluid is defined by its viscosity. The flow is defined by its velocity, vorticity, and turbulence. The regime can be defined by its boundaries.

Under the conditions of a smooth, circular cylinder of infinite length that has been fixed with its axis perpendicular to the direction of a uniform flow without turbulence and with no boundary effects, the fluid-structure interaction can be uniquely defined by the Reynolds number,

$$\text{Re} = UD/\nu, \tag{1}$$

where U = fluid velocity, D = cylinder diameter, and ν = fluid kinematic viscosity. Under these stringent conditions, the fluid-structure interaction can be subdivided into four distinct flow regimes (Le Méhauté, 1976) which are described in Table 1.

Table 1. Flow regimes in a uniform flow past a cylinder.

Flow region	Re range	Flow condition	Force on cylinder
Laminar	0–40	No separation of flow from the body. Continuous stream lines.	In-line forces only.
Subcritical	$40-5 \times 10^5$	Broken stream lines. Well-defined periodic vortex shedding.	Strouhal number dependent lift force frequency. Steady and oscillating in-line forces. Oscillating component at twice the lift force frequency.
Supercritical	$5 \times 10^5 - 7 \times 10^5$	Ill-defined vortices.	Rapid decrease in drag. Lift and drag forces at higher frequencies.
Transcritical	$> 7 \times 10^5$	Vortices are less well defined and persistent. Turbulence due to fluid viscosity produces a randomness factor.	Fluid-structure interaction similar to subcritical range.

The numerical values of the Reynolds number regimes presented in the table should not be considered inviolate. These are estimations of the boundary values. Although the boundaries may have precise values, no test has yet been performed that can be said to completely satisfy the ideal conditions assumed. In other words, no test has yet been performed in which the fluid flow is truly uniform and without turbulence, where the structure is completely fixed, smooth and infinitely long, and where there are no boundary effects. Each of these "real world" parameters affects the fluid-structure interaction and shifts the boundaries of the flow regime. In addition, very few test programs have successfully entered the transcritical regime in a laboratory situation where these parameters can be suitably controlled.

Research programs have been performed which sought to identify the effects of independent nondimensional parameters with respect to the boundary values and subsequent fluid-structure interaction and restraint loads. A comprehensive list of these nondimensional parameters is presented in Table 2. Some of these parameters involve the forced oscillation or free vibration of cylinders. Some parameters may be important for one set of structures but not significant for others. Three nondimensional parameters listed in Table 2 are not considered to be independent variables. They are the drag, lift, and side force coefficients.

1.1. *Fixed structures*

A large body of literature exists on the drag forces experienced by a two-dimensional circular cylinder in a uniform flow field. Laboratory experiments have been performed to determine drag coefficients, and pressure distributions and points of flow separation on the cylinder in uniform flow (Hoerner, 1965). These tests were extended to other two-dimensional bodies, e.g., ships (DHI, 1986). However, many flow fields in nature differ from the idealized uniform flow. It is frequently found that the current near the free surface is stronger than near the ocean floor. A linear shear is often a more realistic situation than uniform flow. Common causes of shear profiles include wind-driven circulations, tidal flows, and bottom boundary effects in shallow water.

It has been shown that a shear current produces lateral force coefficients on a tanker (Palo, 1986) which differ significantly from the force coefficients in uniform current. Because vortex shedding from the bluff body may cause structural oscillation, vortex shedding phenomena in a shear flow should be identified.

Table 2. Summary of nondimensional variables.

Nondimensional variable	Symbol	Definition	Parameters
Reynolds number	Re	$\dfrac{U_M D}{\nu}$	U_M = mean velocity
		$\dfrac{U_M T}{\nu}$	T = ship draft
		$\dfrac{U_M B}{\nu}$	B = ship beam
Strouhal number	S_o	$\dfrac{f_o D}{U_M}$	f_o = shedding frequency
	S_c	$\dfrac{f_c D}{U_M}$	f_c = frequency of forced oscillation
	S_s	$\dfrac{f_s D}{U_M}$	f_s = actual shedding frequency in forced vibration
	S_e	$\dfrac{f_o D}{U}$	U = local velocity in shear flow
Frequency ratio	f^*	$\dfrac{f_c}{f_s}$	
Shear parameter	β	$\dfrac{\lambda D}{U_M}$	
	λ	$\dfrac{dU}{dy}$	y = vertical axis coordinate
Aspect ratio	R_A	$\dfrac{L}{D}$	L = cylinder length
Blockage ratio	R_B	$\dfrac{LD}{WH}$	W = chamber width
			H = chamber height
Depth ratio	R_D	$\dfrac{H}{T}$	
Roughness parameter	k_s	$\dfrac{k}{D}$	k = surface roughness
Reduced velocity	V_r	$\dfrac{U_M}{f_n D}$	f_n = nature frequency of structure
Stability parameter	K_s	$\dfrac{2 M_e \delta}{\rho D^2}$	M_e = total mass (including added mass) per foot
			δ = logrithmic decrement in air
			ρ = fluid density

Table 2. (*Continued*)

Nondimensional Variable	Symbol	Definition	Parameters
Free amplitude ratio	R_y	$\dfrac{y_o}{D}$	y_o = amplitude of free oscillation
Forced amplitude ratio	R_a	$\dfrac{a}{D}$	a = amplitude of forced oscillation
Elevation ratio	R_E	$\dfrac{y}{L}$	
Turbulence intensity	T_I	$\dfrac{U_{dev}}{U}$	U_{dev} = deviation of velocity from mean velocity
Drag coefficient	C_{D_f}	$\dfrac{F_{I_f}}{1/2\rho U^2 D\Delta L}$	F_{I_f} = in-line force at specified frequency, f
Lift coefficient	C_{L_f}	$\dfrac{F_{L_f}}{1/2\rho U^2 D\Delta L}$	F_{L_f} = transverse force at a specified frequency f
			ΔL = length over which force is measured
Lateral force coefficient	C_y	$\dfrac{F_M}{P_d L_{pp} T}$	F_M = measured force on ship beam
			p_d = dynamic pressure
			L_{pp} = length between perpendiculars
Pressure coefficient	C_p	$\left[\dfrac{P - P_o}{1/2\rho U^2}\right] + 1$	p_o = stagnation pressure
			p = pressure around cylinder circumference
Velocity gradient	ε	$\dfrac{U_{MAX}}{U_M} - 1$	U_{MAX} = maximum velocity
Energy coefficient	α	$\dfrac{\int_o^H U(y)^3 dy}{U_m^3 H}$ $= 1 + 3\varepsilon^2 - 2\varepsilon^3$	
Surface effect coefficient	C_s	$\dfrac{y_1}{D}$	y_1 = vertical distance from cylinder longitudinal axis to free surface
Bottom effect coefficient	C_b	$\dfrac{y_2}{D}$	y_2 = vertical distance from cylinder longitudinal axis to bottom boundary

1.2. Floating structures

It is a common practice to moor tankers and barges in channels with high current flows. Several mooring lines are used to prevent these vessels from drifting. In order to design these mooring lines, the current load on the vessels should be known. Analytical methods to compute these forces are still not adequate. Therefore, they are determined from controlled laboratory experiments and field tests. OCIMF (1977) has extensive literature on the prediction of current loads on two types of bow configurations for large crude carriers. The data for these curves generally came from model tests which have included fixed and moored models of vessels in current as well as towed models in still water. However, the validity of some of these curves has been questioned, e.g., (Palo, 1986).

Recently, several researchers have reinvestigated the effect of current on floating vessels (Fellmann, 1992). Current forces on a large scale model were reported by Edwards (1985). The effect of proximity of two large vessels on current forces were examined by DHI (1986), Jacobsen (1987), and Kwok (1989). Kriebel (1992) carried out experiments on FFG-7 and Series 60 hulls. These data and others were presented to show tank blockage effects by Seelig et al. (1992) who developed a blockage coefficient. Chakrabarti and Cotter (1993) further investigated the tank wall effect on moored barges in current. While all of these tests deal with uniform flow, current in nature is often found to be stronger near the surface than at the bottom. A shear current produces a significantly different lateral force on a vessel than the corresponding uniform flow.

NCEL carried out a full-scale test (Palo, 1986) of a T-2 tanker moored in steady current. The results indicated that the positive shear (decreasing values with depth) tends to reduce the lateral load compared to the load in a uniform flow field while the negative shear (increasing values with depth) increases the lateral load. While there was considerable scatter in the data, the shallow water effect on the lateral force coefficient showed a possible increase with Reynolds number, which was absent in the deep water results.

2. Shear Current Generation

It is often desired to duplicate experimentally the nonuniform flow in a flow field that is larger than the normal frictional flow developed in any generated flow field. This flow field is usually generated just upstream of the test section.

This is most easily accomplished by placing a disturbance in the flow upstream where the fluid is otherwise uniform (and, in fact, potential except for thin boundary layer regions).

The selection of the disturbance is important such that it does not greatly influence the downstream flow in the test section. The requirements of an artificial shear flow should include the following (Kotansky, 1966):

(1) steady flow downstream of the disturbance;
(2) smooth variation of velocity with a minimum of local small scale nonuniformites; and
(3) negligible diffusion of the shear layer through the length of the test section.

The total pressure loss through the device should be minimized so that the operating characteristics of the flow generation remain relatively unaffected.

A shear flow can be generated in flowing air or water in a number of ways. The boundary layer provides a shear flow. Mechanical means such as curved screen, array of rods or screens (or mesh) are also used. In a wind tunnel, the boundary layers are often thickened with the help of gauges or honeycombs, bank of rods, or a grid of flat plates. Other methods used in air or water are a plane rectangular mesh or circular wires, a sheet of cloth, a cascade of airofoils, a honeycomb of parallel tubes, or a combination of these methods (Elder, 1959).

The use of stacked thicknesses may result in poor velocity profiles. An open cell aluminium honeycomb was found to be very good in generating shear flow. Unlike many other materials, it usually does not require additional structural support. The honeycomb may be easily varied to obtain a wide range of flow geometries; it has a built-in flow straightener; the structure of the flow downstream is homogenous.

The ability to realistically model a turbulent shear flow at high Reynolds number past a structure (even a cylinder) is beyond the capabilities of existing numerical fluid dynamics algorithms, even though attempts are being successfully made in approaching this physical situation. The results on shear flow effect thus far are mostly experimental.

2.1. Current map

The shear fraction is given in terms of the ratio of the change in the velocity profile across the shear (ΔU) and the free stream velocity (U_∞). The shear strength is defined by the nondimensional quantity β as

$$\beta = \frac{\partial U}{\partial y} \frac{D}{U_{\max}}, \qquad (2)$$

where the first term on the right-hand side is the slope of the current profile with depth (velocity gradient) while the second term is used to nondimensionalize the quantity. The numerator represents a transverse dimension to flow (e.g., cylinder diameter, vessel draft, etc.) while the denominator is taken as a reference fluid velocity at the structure. The reference velocity is sometimes considered as the free stream velocity before it reaches the shear screen. In Eq. (2), the velocity represents the maximum time-averaged current amplitude produced at the structure center line by the facility for a given test set-up. Some investigators use local velocity and others use mean velocity. For a floating structure, sometimes twice the draft is used for the transverse dimension on the assumption that the vessel represents a double-body at the water surface. Since $U_{\max}$ is used for all cases of current generation in Eq. (2), the shear strength is strictly a function of the slope of the profile. While this definition is consistent within the same scale and follows Froude's law, it is somewhat artificial by the choice of $U_{\max}$ irrespective of the test current generated. Note that this definition of shear (Eq. (2)) is consistent with the definition used by Chakrabarti, Cotter, and Palo (1993) but differs from the definition given by Palo (1986). The definition given by Palo (1986) differs by the choice of a particular velocity in the denominator (namely, mid-draft elevation) based on a particular test case. This definition may provide a higher normalized vertical shear (NVS) value for a lower slope if the current value at the vessel for the test set under consideration is small.

It seems that no definition for the shear strength can be found to be satisfactory, even if it is consistent within a given study. Therefore, care should be taken when a comparison is made between two separate studies or when data is scaled up to prototype values.

The current profile is generally mapped at the test site before any test commences. The current speed is recorded with a moving current meter or a series of meters simultaneously. In a test, shear currents were generated by providing increasing levels of friction to flow at prescribed elevations. This friction was achieved by installing a mesh in front of the flow straighteners placed across the flow cross-section. The flow straighteners consisted of 6-inch diameter, 4-feet long PVC tubes. The mesh consisted of a polyethylene hardware cloth with a diamond-shaped mesh pattern. The cloth commercially available in 36-inch-wide rolls with nominal openings of 1/8", 3/16", 1/4", 1/2", 3/4" and 1".

The cloth was cut into 6-inch widths and various combinations of mesh openings were tried at each elevation. Added resistance was provided to the flow by this cloth and the subsequent shear pattern was developed by a trial and error approach.

The mesh was then turned upside down to produce the negative shear. Each mesh produced a shear parameter of $\beta = \pm .06$ between the elevation of 12" and 24" from the floor.

The current was mapped for uniform, positive shear and negative shear cases for both deep and shallow water. The profiles showing the time-averaged mean current speed at each measurement location for the shallow water case are shown in Figs. 1–3. The profiles shown in each figure represent different

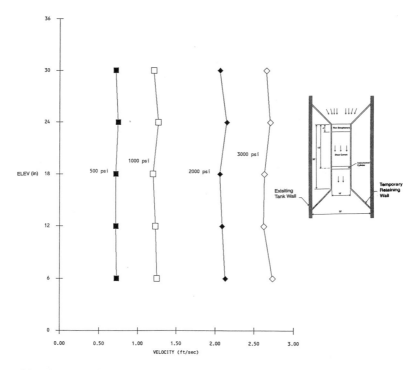

Fig. 1. Mapping of uniform fluid flow of different strengths in the constricted area shown as insert. The pressure settings (psi) of the current generator are shown.

240 S. K. Chakrabarti

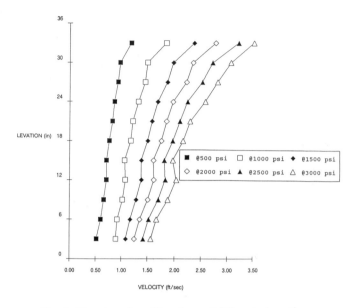

Fig. 2. Mapping of positive shear of different strengths.

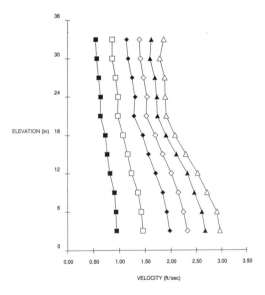

Fig. 3. Mapping of negative shear of different strengths. (See Fig. 2 for explanation.)

current strengths and show a similar trend. According to the definition in Eq. (2), the shear strength (positive or negative) is different for each current setting shown since the slopes of the vertical profiles are different, but the maximum reference velocity, $U_{\max}$, is the same for all profiles.

3. Effect of Fluid Flow on a Vertical Cylinder

Vertical cylinders, as members of an underwater support structure, experience loading from the flowing fluid. This section discusses the differences in the fluid-structure interaction for a vertical cylinder between a uniform fluid flow and a vertical shear flow.

3.1. *Vertical cylinder in uniform flow*

For an infinitely long, smooth, fixed, vertical cylinder in a uniform flow without turbulence, the fluid-structure is defined by the Reynolds number (Hoerner, 1965). Thus, for a constant viscosity fluid, the fluid-structure interaction remains the same, and the value of the steady (in-line) drag coefficient remains the same if the velocity is doubled at the same time the cylinder diameter is decreased by a factor of two.

If the undisturbed flow is in the subcritical range, the vortices will continue to be well defined and regular, and the shedding frequency will be characterized by the Strouhal number. In the subcritical regime, the Strouhal number is about 0.2. Within the transcritical regime, the Strouhal number is approximately 0.3. In the supercritical transition region, regular vortex shedding ceases, and the Strouhal number becomes ill-defined (Humphries and Walker, 1987). The cylinder roughness and fluid turbulence tend to lower the boundary value on either side of the supercritical region. The decrease is more pronounced on the upper boundary, and the supercritical region tends to become narrower.

End effects introduce the aspect of three-dimensionality to the flow. The ends of the cylinder affect the pattern of vortex shedding and the values of drag and its coefficients. Vortices are not shed uniformly along the cylinder length, and the drag and lift coefficients become dependent on the location along the cylinder. For all aspect ratios up to $L/D = 20$, the vortex shedding pattern at the center of the cylinder is affected by the ends. This is illustrated in Fig. 4 (Mair and Stansby, 1975), the data of which was obtained in a shear flow.

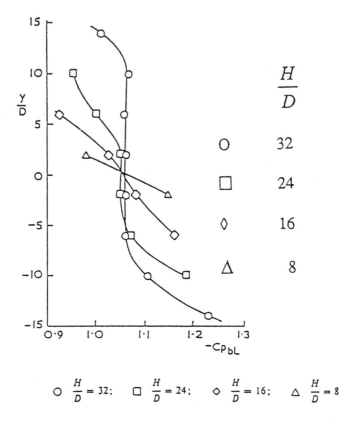

Fig. 4. Base pressure on vertical circular cylinder as a function of aspect ratio and distance from ends. (Mair and Stansby, 1975)

The cylinder often has some freedom of movement because of its inherent flexibility or the flexibility of its mounting apparatus. For ease of control, many investigators have employed a test set-up of forced oscillations to duplicate this flexibility. In this case, reduced velocity V_r, an additional nondimensional term (Table 2), becomes important. Reduced velocity characterizes vortex-induced vibrations by different instability regions (Humphries and Walker, 1987) which have been summarized in Table 3.

Within these three stability regions, the stationary cylinder Strouhal number is no longer valid, and vortex shedding takes place at the structural vibrational frequency. This phenomenon is commonly known as lock-on. The

Table 3. Vortex induced vibration of a cylinder. (Humpries and Walker, 1987).

Region	Reduced velocity	Vortex shedding	Oscillation
I	1.7–2.3	Symmetrical shedding	In-line oscillation
II	2.8–3.2	Alternate shedding of vortices from cylinder	Predominantly in-line vibration
III	4.5–8.0	Alternate vortex shedding	Predominantly transverse vibration. In-line vibration at twice the transverse vibration frequency, resulting in a figure eight motion.

boundaries of these instability regions are affected by the stability parameter, end effects, boundary effects, and turbulence. Stansby (1976) showed that the range of lock-on increases with the amplitude of the cylinder motion. He also demonstrated that lock-on can occur at one-half or one-third of the structural frequency.

3.2. Vertical cylinder in shear flow

When a vertical cylinder is in a shear flow, a three-dimensional flow regime will occur. For a positive shear parameter (greatest velocity near the surface), the variation of stagnation pressure on the upstream face of the cylinder promotes a vertical flow down the face of the cylinder. Conversely, the varying pressure in the wake of the cylinder promotes a vertical flow up the rear of the cylinder (Fig. 5).

These three-dimensional flow cells tend to reduce the expected dynamic pressure coefficient at the top of the cylinder and increases the coefficient at its base (Fig. 6).

In the absence of end and boundary effects, a vertical cylinder in uniform flow will shed vortices at the same frequency, as defined by the Strouhal number, over the entire length of the cylinder. In a shear flow, the vertical cylinder is expected to shed vortices at a Strouhal frequency that is changing continuously with the velocity and a continuous spectrum of frequencies would arise. However, in a shear flow on a vertical cylinder, vortex cells are formed over which the shedding frequency does not vary continuously (Fig. 7). The

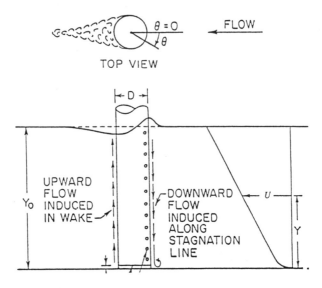

Fig. 5. Schematic of change of stagnation pressure on vertical cylinder in shear flow. (Masch and Moore, 1960)

number of cells and the extend of each cell are dependent on the aspect ratio, end and boundary effects, surface roughness, and turbulence. There is a general trend toward decreasing cell length with increasing shear and cell length with cylinder roughness (Griffin, 1985).

In a test with a vertical cylinder with a relatively small aspect ratio ($L/D = 8$) in a shear flow (Mair and Stansby, 1975), only two stable end cells were formed. The bottom boundary layer carried vorticity which augments the shear, while the top boundary (if fixed) carried vorticity in the boundary layer of the opposite sign. For a longer cylinder ($L/D = 16$), there was a region in mid-span extending about 8D which had cells distinct from the end cells, even though the cell structure was still dependent on the end conditions. Typically, the longest cells were in the order of 4D to 6D.

At the intersection of the cells, there is a region of instability in which the shedding frequency tends to alternate between cells. The addition of end plates stabilizes these regions and makes the cell structure more distinct. In order to stabilize the cells, various investigators tried end plate configurations in which the end plate circumference was an integer multiple of the cylinder

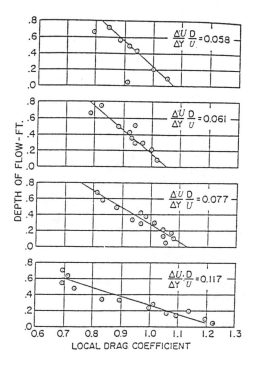

Fig. 6. Variation of local drag coefficient with shear parameter. (Masch and Moore, 1960)

diameter. The optimum integer value appeared to be a function of the flow velocity. These plates, however, simply stabilize the cells; they do not remove the end effects.

In order to display the cellular structure of the vortices in shear flow, the Strouhal number is defined in terms of the local velocity or the mean velocity. If the local velocity is used in the definition, there will be a continuous change in the Strouhal number across the cell in order to maintain a constant frequency, and then a jump in value will occur at the discontinuous cell change. Using the mean velocity, U_M, the Strouhal number (Fig. 7) will remain a constant over the cell length and then jump to a new constant value at the next cellular region (Maull and Young, 1974). In either case, the Strouhal number cannot be considered as an independent variable in shear flow on a vertical cylinder and can only be used as an indication of the frequency range of vortex shedding that may exist.

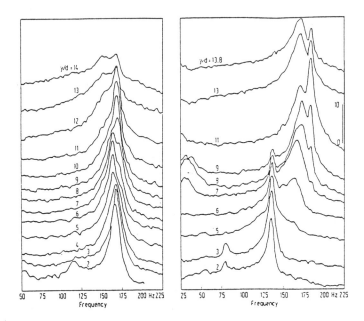

Fig. 7. Frequency content of wake behind vertical cylinder in uniform flow (left) and shear flow (right). (Maull and Young, 1974)

In the same way, the shear parameter, β, is somewhat suspect as an independent variable. Apparently, no data exists demonstrating identical cellular flow patterns for test set-ups with twice the diameter and half the shear. In fact, inspection of the Strouhal number would indicate that the cellular pattern would change.

4. Effect of Fluid Flow on a Horizontal Cylinder

Horizontal cylinders, such as subsea pipelines, encounter current in riverbeds and coastal regions. The stability of pipelines and their foundation design depends on the loading imposed on them by the fluid flow past them. Horizontal cylinders are also found in subsurface and near surface regions as members of offshore structures, pontoons, etc. This section describes the effect of uniform and shear current flow on a horizontal cylinder placed across the flow.

4.1. *Horizontal cylinder in uniform flow*

Boundary effects and end effects on a horizontal cylinder introduce three dimensionality. In a uniform flow in the absence of a boundary, the end effects are considered to be identical between a horizontal and a vertical cylinder.

Near the upper boundary, i.e., the free surface, the wave resistance of a horizontal circular cylinder in a uniform velocity field is dependent on the Froude number: $F_r = U^2/gy_1$ where y_1 is the depth of immersion measured from the cylinder axis to the free surface and g is the gravitational acceleration. When y_1 is large compared to the cylinder radius, the maximum wave resistance occurs for $F_r = 1$ (Bishop and Hassan, 1964). The wave resistance is negligibly small when $F_r < 0.375$. For towed ship shapes, the wave resistance drag is quite small for $F_r < 0.2$ (Hoerner, 1965), and the residual drag (i.e., form plus wave drag) is not a function of the Froude number for Froude numbers less than 0.5. Near the bottom boundary (or fixed top boundary), the flow becomes retarded, and a shear flow is developed. Thus, the vortex shedding pattern is affected by the proximity of the boundary.

The major difference in the boundary effects between a horizontal and a vertical cylinder is the appearance of a lift force. As the horizontal cylinder approaches either boundary, the flow can no longer be symmetric, and the lift force becomes a function of the cylinder diameter and the distance to the boundary wall. A horizontal cylinder in the uniform flow near a plane boundary can be treated as a pair of horizontal cylinders in the absence of a boundary. Data on the drag coefficient and Strouhal number as a function of the separation distance between the two cylinders (Hoerner, 1965) suggest that the proximity effect is negligible for gap/diameter $(y_2/D) > 4.0$.

4.2. *Horizontal cylinder in shear flow*

When the cylinder is mounted horizontally in a shear flow, the relationship between the diameter, D, and the shear, dU/dy, takes on a physical significance in terms of velocity variation across the face of the structure. In this case, the mean velocity and the local velocity nearly coincide. In the absence of boundary or end effects, the vortex cell structure exhibited by the vertical cylinder should disappear and the Strouhal number should regain its significance. However, the ratio of turbulence to velocity variation across a horizontal cylinder will be much larger than for a vertical cylinder. It is important to recognize the importance of turbulence when comparing the results of shear flow to uniform flow. Ideally, for a one-to-one correspondence, both flows would have similar turbulence intensities.

A test on a horizontal cylinder in a shear flow was carried out by Kiya et al. (1980). The wake characteristics behind the cylinder were measured as a function of the shear parameter and the Reynolds number. The aspect ratio in their tests varied from 2.0 to 12.5. In the uniform flow, the transition into vortex shedding occurred at a slightly higher Reynolds number (= 52) than the generally accepted value of 40. This increase in the transition level was attributed to an (unmeasured) increased level of turbulence. The shear parameter ranged from 0.05 to 0.25. As the shear parameter increased, the Strouhal number, and hence the vortex shedding frequency, was found to increase (Fig. 8). Similarly, the transition Reynolds number at which vortex shedding appears increased with the increase in the shear parameter (Fig. 9).

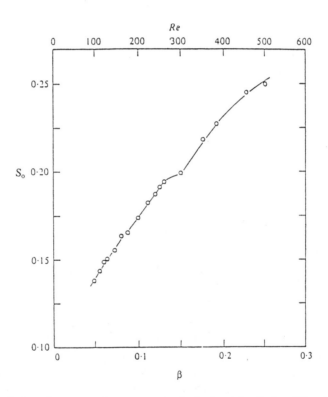

Fig. 8. Strouhal number versus shear parameter for horizontal cylinder. (Kiya, Timura and Arie, 1980)

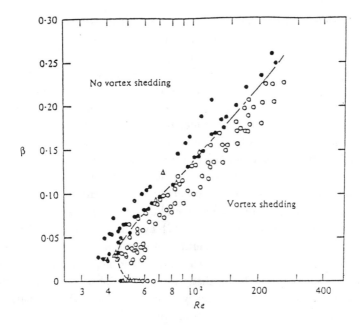

Fig. 9. Vortex shedding regions versus Reynolds number and shear parameter (●—steady twin vortices; ○—vortex shedding). (Kiya, Timura and Aire, 1980).

4.3. *Test Results from a Horizontal Cylinder*

In a test with a fixed horizontal cylinder, both uniform and shear current flows were generated (Chakrabarti et al., 1993). This allowed the comparison of loads in these flows and demonstrated the special effect of shear flow on the loads.

The model was a smooth, 6-inch diameter, aluminium cylinder. It was 10 feet long and was fixed at either end to the restraining wall, 18", 24" and 33" above the basin floor respectively in a 36-inch water depth. At the cylinder mid-span was a one-foot long instrumented section capable of measuring the local load on the cylinder in two orthogonal directions. A pluck test demonstrated that the model's first mode of vibration was 2.14 Hz which was separated far from the computed Strouhal frequency of about 1 Hz for this test set-up.

4.3.1. *Steady loads*

When the cylinder is at mid-depth, the positive and negative shear currents create the same load on the cylinder, and this load is larger than the load created by the uniform flow at the same velocity. When the cylinder is near the surface, the uniform current produces a larger load than the positive shear for the same velocity. The negative shear did not provide a large enough velocity at the surface to make a reasonable comparison. At an intermediate elevation, the negative shear creates a larger load than the uniform flow or positive shear. As the velocity becomes larger, the positive shear appears to produce higher loads compared to the loads from the uniform flow. In general, loads are relatively independent of elevation in positive and negative shear and depend only on velocity. In uniform current, however, the in-line load on the cylinder increases as the cylinder approaches the free surface.

The steady in-line load on the horizontal cylinder is presented as a function of velocity and shear pattern for the 18-inch cylinder elevation in Fig. 10. The mesh pattern used for the shear generated is shown as an inset in the figure.

The same pattern was inverted to produce the negative shear. The positive shear produced larger loads than the negative shear or uniform current which produced similar loads. As the cylinder approached the free surface, the vertical load in all shears increased dramatically (Fig. 11).

4.3.2. Dynamic component of vertical load

The vertical loads on the cylinder were oscillatory. Frequency domain analysis showed that at lower velocities, the vortex shedding was highly defined in terms of a single frequency. However, as the velocity increased, the flow became more turbulent. This is illustrated for the mid-depth (18 inches) in Fig. 12 for negative shear. The spectra are arranged at increasing values of the mean velocity for the negative shear. The shear strength (based on Eq. 2) and Strouhal numbers are shown in each case. At the higher velocities, Strouhal numbers are undefined.

The dynamic vertical loads on the cylinder at mid-depth (18 inches) in positive and negative shears contain greater energy at the higher frequencies and overall contain 30–40% more energy than the load in uniform current. Near the surface (33 inches), no strong trend between shear types is discerned, but there is a marked reduction (about 50%) in the significant value of the energy in comparison to the data at mid-depth. In this case, the surface tends to suppress the eddy shedding and wake formation. It should be remembered, however, that a large steady vertical load arises as the cylinder approaches the surface.

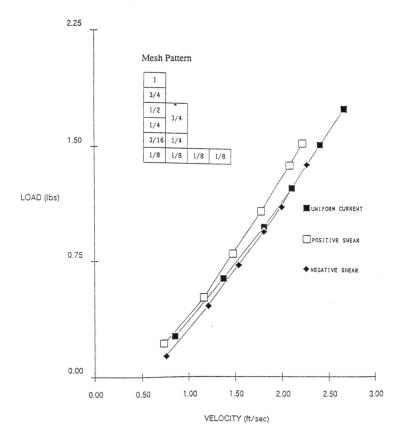

Fig. 10. In-line load on horizontal cylinder at an elevation of 18 inches.

The frequency of the periodic velocity fluctuations in the wake of the cylinder is represented by the Strouhal number. Strouhal numbers were calculated for the cylinder at mid-depth and velocities below 1.8 ft./sec. based on the spectral peak frequencies. These values are shown in Table 4 for different flow fields. Kiya *et al.* (1980) presented values of the Strouhal number in the Reynolds number range, $35 < \text{Re} < 1500$, in which the Strouhal number was found to increase with shear parameter. The values presented in Table 4 for shear parameters in the range of 0.08 to 0.20 compare with the results of Kiya *et al.* A slight tendency for the Strouhal number to increase from uniform to shear flow is also indicated.

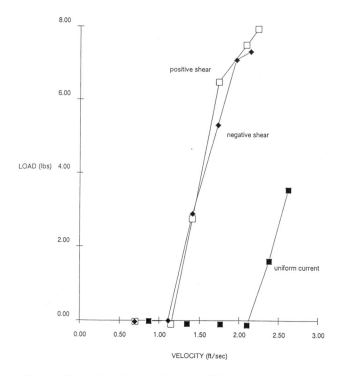

Fig. 11. Vertical load versus shear condition at the free surface.

Highly defined vortex shedding is apparent at mid-depth at the lower Reynolds number. The Strouhal number in this case tends to increase with the magnitude of the shear parameter. At a higher Reynolds number and near the surface, the flow is more turbulent and the Strouhal number is less well defined.

4.3.3. *Pressure profile around cylinder*

The pressure profiles around the cylinder at mid-depth were measured for three types of flow: uniform current, positive shear, and negative shear. All three profiles are found to be symmetric with the positive shear displaying a slightly larger magnitude on the top of the cylinder (90°), and the negative shear displaying a slightly larger magnitude at the bottom of the cylinder (270°). Figure 13 presents similar data for the cylinder at the surface. The symmetry in the mean pressure profile disappears as the free surface is approached.

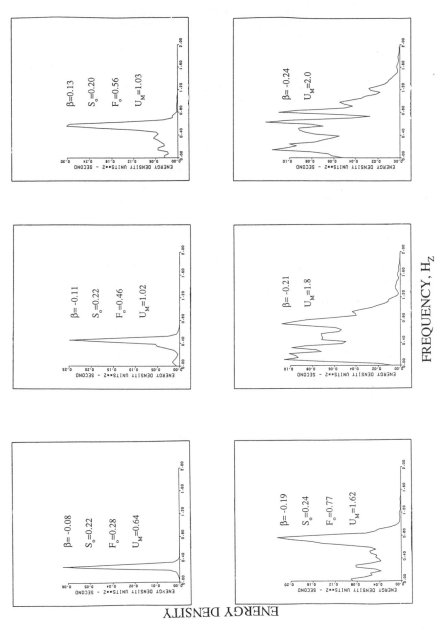

Fig. 12. Transverse (vertical) load spectrum of horizontal cylinder at mid-depth; spectral peak represents shedding frequency.

Table 4. Strouhal numbers for various shear parameters.

Shear	Velocity (ft./sec.)	β	Re no. $\times 10^4$	Peak freq. (H$_z$)	St. no.
Uniform	0.82	0	4.10	.35	.21
	1.41	0	7.05	.60	.21
	1.80	0	9.00	.74	.21
Positive	0.64	0.09	3.20	.28	.22
	1.02	0.13	5.10	.46	.22
	1.57	0.15	7.85	.70	.22
	1.66	0.20	8.30	.67	.20
Negative	0.64	-0.08	3.20	.28	.22
	1.03	-0.11	5.15	.46	.22
	1.38	-0.13	6.90	.56	.20
	1.62	-0.19	8.10	.77	.24

The upper forward quadrant ($0° < 0 < 90°$) tends to have a positive pressure and the lower forward surface quadrant ($270° < 0 < 306°$) tends to have a larger negative pressure.

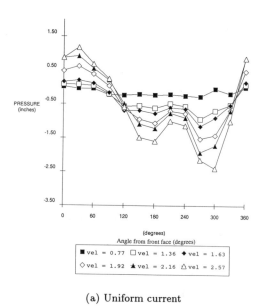

(a) Uniform current

Fig. 13. Pressure profile around cylinder near free surface in current.

Shear Current and Its Effect on ... 255

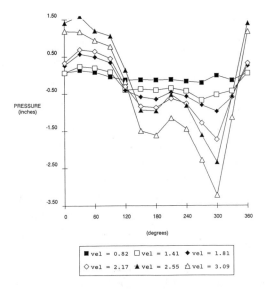

(b) Positive shear

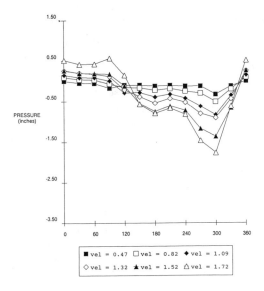

(c) Negative shear

Fig. 13 (*Continued*).

4.3.4. Drag coefficient

The drag coefficient for the cylinder corresponding to the in-line load was analyzed versus the Reynolds number at the three elevations. When the cylinder was at mid-depth in uniform flow, the drag coefficient was near the expected established value ($C_D \approx 1.1$) for low values of the Reynolds number (Re= 4.4×10^4). As the Reynolds number increased (Re = 1.25×10^5), the drag coefficient decreased rapidly to a value of 0.5 which is normally associated with the fully developed transition zone (Re $\approx 6.0 \times 10^5$). This "early" entrance into the transition zone may be caused by the presence of turbulence in the flow. While turbulence was observed near the surface and likewise expected at mid-depth, the velocity measurement showed no clear evidence of turbulence mainly because of a built-in filter in the probe to reduce inherent electrical noise in the data.

The drag coefficient is higher in shear flows than in uniform flows when the cylinder is at mid-depth. All values appear to approach $C_D = 0.5$ as the transition zone is entered. When the cylinder is at intermediate depth, the drag coefficient is the largest in negative shear. When the cylinder is near the surface (Fig. 14), the value for C_D in uniform flow is greater than the value in positive shear. Again, the negative shear tends to have the highest values at the low Reynolds number range (Fig. 14).

5. Shear Flow-Induced Vibrations

The resonant flow-induced vibration of a stationary structure occurs when the vortex shedding frequency due to the flow of fluid past the structure approaches one of the natural frequencies of the structure. This is known as "lock-on" or capture of the vortex shedding frequency by the vibration frequency over a range of flow speeds. This lock-on effect causes the wake and the structure to oscillate in unison (Griffin, 1985). When the flow is nonuniform, there is an added effect of a vorticity in the approaching flow. The vorticity interacts with the vortices which are shed from the body into its wake. For a vertical cylinder in shear flow, the ratio of length-to-diameter ratio, L/D, is an important parameter, in addition to the steepness or shear (strength) parameter.

In a highly turbulent shear flow ($\beta = 0.18$), the critical Reynolds number has been observed to be reduced by a factor of ten. In a low and moderate-turbulent shear flow, the vortex shedding pattern is cellular. The cell length has been found to decrease with increasing shear. For a long flexible cable ($L/D = 100$) a discernable cell structure existed at moderate subcritical

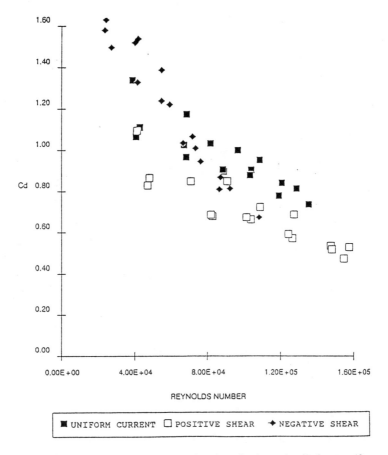

Fig. 14. Drag coefficient versus Reynolds number for a horizontal cylinder near free surface in uniform and shear-flow.

Reynolds number for $\beta = 0.005$ (Peltzer, 1982). Davies (1976) studied the effect of a highly turbulent shear flow on the vortex shedding from a vertical circular cylinder. The shear parameter value was $\beta = 0.18$ and the turbulence level (U_{dev}/U) was 5%. The base pressure coefficient at the mid-span of the cylinder is shown in Fig. 15 for both uniform and shear flows. It is clear that the onset of the critical Reynolds number is reduced in the shear flow by a factor from the uniform smooth flow value of Re $= 2 - 3 \times 10^5$. The reduction in Re_{cr} is similar to that produced by a uniform flow of the same turbulence intensity.

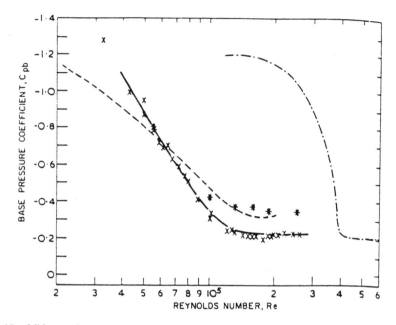

Fig. 15. Mid-span base pressure coefficient C_{pbm} for a circular cylinder in a shear flow (Davies, 1976). ×, centerline value ($\beta = 0.18$); ..., uniform smooth flow; − − −−, uniform turbulent flow; ∗, average value ($\beta = 0.18$).

The cell structure at the subcritical Reynolds number is well defined with strong vortex shedding and associated predominant wake frequency. As the Reynolds number increased, the shedding patterns became irregular, disappearing into a turbulent background at Re = 10^5. The distribution of the base pressure coefficient, C_{pb}, measured on a vertical cylinder in a linear shear flow is shown in Fig. 16 (Peltzer and Rooney, 1980–81). In this case, the length ratio $L/D = 48$, roughness $k/D = 0.001$, $\beta = 0.015$, and Re = 2×10^4. This result is typical showing the end effect adjacent to the end plates of the cylinder. In general, the vortex shedding patterns were free of constant frequency cells.

A cellular pattern of vortex shedding exists along the span of a stationary vertical cylinder in a shear flow. Over each cell the vortex shedding frequency is constant. The Strouhal number and base pressure coefficient are also found to be constant. The experimental results are generally limited by the small cylinder aspect ratio ($L/D < 20$).

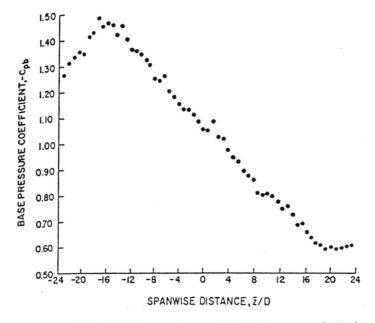

Fig. 16. Spanwise variation of the base pressure coefficient C_{pb} on a rough ciruclar cylinder in a shear flow; for $Re = 2 \times 10^4$ and $\beta = 0.015$. (Pelzer and Rooney, 1980)

6. Effect of Fluid Flow on a Moored Floating Structure

A ship is often moored in a strong current in the general absence of waves. This is particularly true for ships or barges in a narrow channel. The floating structure in this case is allowed to move in a current within the constraints of the mooring system. The mooring line design depends on the knowledge of the effect of uniform and shear currents on the floating structure.

6.1. *Possible implications of shear flow to mooring system design*

A ship moored in a beam current represents a blunt object in a fluid flow field. Even in its most generalized conceptualization as a half cylinder at the free surface, it cannot be represented as a baseline condition because of the influence of the free surface. The addition of keel configurations, stern and bow effects, seafloor effects, ship motion, and shear current creates an increasingly complex flow regime.

Nevertheless, it can be assumed that the lessons learned from experiments with horizontal and vertical cylinders in uniform and shear flow can be used as a foundation for understanding this flow regime. The moored vessel will similarly experience four distinct flow regimes as the current (Reynolds number) is increased and the steady drag coefficient will undergo changes as the various regimes are entered. Because the flow is by necessity asymmetric, a steady lift force may exist in all regimes. The moored vessel will experience lift and drag loads at frequencies defined by a Strouhal number when the flow is in the critical or transcritical regimes. However, the wave-making resistance will be more important because the vessel is at the free surface. The lift forces will generate a heave motion on a symmetric vessel and heave, pitch and roll motions on an asymmetric vessel (stern versus bow) with a keel. The frequency of the vortices created by the fluid flow will be different from the frequency of the vortices generated by the ship motion for certain ranges of the reduced velocity and identical (lock-on) at other ranges. Nondimensional parameters which influence the regime boundaries for a cylinder (turbulence surface roughness, bottom boundary effects, shear parameter) will also influence the boundaries of the flow regime of the moored vessel.

As the frequencies of the applied force approaches the natural frequency of the system, the resulting motions and ensuing line loads become amplified and are restricted only by the damping in the system. Experience in offshore mooring systems has taught designers that the applied load does not have to occur at the natural frequency for the system to undergo dynamic amplification. Even in the absence of waves, a similar occurrence will arise if the vortices shed by the fluid passing the hull are at a different frequency than the vortices shed by the ship motion. An oscillating in-line load (drag force) will be associated at twice the value of each frequency of vortex shedding. Even though these loads are quite small (typically the drag coefficient associated with this oscillating in-line load is in the order of 0.2), dynamic amplification could result in a large response.

In the critical region, there are well-defined periodic shedding frequencies. The transcritical region is similar to the critical region except that the vortices are less well defined and persistent. It may be suggested that the lessons learned from experiments in the critical region can be applied directly to the transcritical region where few experiments have been performed and most mooring systems exist. Table 5 presents the Reynolds number ranges for the experiments encountered in the literature. Inspection of this table indicates that very few experiments have even approached the supercritical range.

Table 5. Range of Reynolds numbers tested.

Investigator	Definition	Medium	Range
Stansby (1976)	$U_M D/\nu$	Air	$1.3 \times 10^4 - 2.6 \times 10^4$
Mair & Stansby (1975)	$U_M D/\nu$	Air	$1.3 \times 10^4 - 2.6 \times 10^4$
Maull & Young (1974)	$U_M D/\nu$	Air	2.85×10^4
Masch & Moore (1960)	$U_M D/\nu$	H_2O	$1.85 \times 10^4 - 2.25 \times 10^4$
Zedan, Seif & Shibl (1988)	$U_M D/\nu$	Air	5×10^5
Bishop & Hassan (1964)	$U_M D/\nu$	H_2O	$4 \times 10^3 - 12 \times 10^3$
Kiya (1988)	$U_M D/\nu$	H_2O	$35 - 1500$
Rooney & Peltzer (1981)	$U_M D/\nu$	Air	$2.5 \times 10^5 - 5.2 \times 10^5$
Humphries & Walker (1987)	$U_M D/\nu$	H_2O	$0 - 2.4 \times 10^5$
Grosenbaugh & Yoerger (1989)	$U_M D/\nu$	H_2O	$10,000$
Shaw & Starr (1972)*	$U_M D/\nu$	H_2O	1.3×10^4
Edwards (1985)	$U_M B/\nu$	H_2O	7×10^5
Palo (1983)	$U_M (2T)/\nu$	H_2O	7×10^5

*used glycerin to modify the water viscosity

Edwards (1985) suggests that the keel on his model tanker created an early entrance into the transcritical region. During this test program, the effect of the depth parameter was far less pronounced than in previous tests performed in the critical region.

This raises the question as to the applicability of data generated in the critical region to the transcritical region. It is certainly easier to perform tests in the critical region. Not only is the size of the experiment more accommodating, but the flow patterns are more defined and persistent. The increased level of turbulence associated with the transcritical region introduces a randomness to the data and makes the data more difficult to process and present in a meaningful method. Therefore, a good test plan may operate generally in the critical region where trends are less obscure, but at some point, tests must be performed in the transcritical regime to validate the application of the trend to the transcritical regime.

6.2. Tank wall effects

Model tests are often performed in tanks of limited width. Therefore, the model dimensions, such as the length in a beam sea test, must be carefully chosen to represent the mooring conditions at sea. Generally, a tank width of 4–5 times the length of the vessel is considered sufficient for beam sea tests.

Jacobsen and Jones (1987) investigated the effect of tank width on the lateral forces on a fixed model in beam seats. For a given draft to depth ratio, (T/d), where T = vessel draft and d = water depth, the ratio of basin width to model length (b/L, where b = basin width and L = model length) was varied from 1.5 to 7.8. The drag coefficient was found to increase with decreasing values of b/L. The effect became negligible when b/L was larger than 5.

In a more recent paper, Seelig et al. (1992) examined the results of a model test for the current forces on moored ships in a beam sea. They derived empirical formulas for the deep water and shallow water drag coefficients. The channel blockage factor, f, was defined as

$$f = \frac{bd}{LD}. \tag{3}$$

Then the correction factor from finite width to the infinite channel was given by

$$C_f = \frac{f-1}{f}. \tag{4}$$

The drag coefficient C_D obtained in a finite width tank was multiplied by C_f to obtain the unblocked channel data.

Chakrabarti and Cotter (1993) tested three barges (large, medium, and small) that had the same form and were dimensionally similar to each other. The basin width (W) was 33 feet while the water depth was varied from 18 to 60 inches resulting in a wide range of d/T values. The three barges were:

Size	Length (ft.)	W/L	Beam (ft.)	Draft (ft.)
Large	11.00	3	1.49	0.69
Medium	6.60	5	0.90	0.41
Small	4.13	8	0.56	0.26

The small barge was modified for certain tests to include the effect of bilge keels.

6.2.1. *Test results*

The results of the test program were presented in terms of drag coefficient C_D versus Reynolds number (Re) for various barge sizes and water depths. The barges were placed in a beam sea condition in the basin. It should be noted

that in a test of this nature, there are at least three separate phenomena that can affect the forces on the barge:
- shallow water effect — tank floor effect,
- side wall effect, and
- three-dimensional flow effect.

The results of the three barge tests are summarized in Fig. 17. All barges were tested without a bilge keel except the small barge. The presence of the bilge substantially increased the C_D value. The results show that a d/T value greater than 4 is sufficient for deep water.

The sensitivity of C_D to Re is small due to the small range of Re. For the medium barge, the mean deep water C_D value is about 0.7 while for the large barge, it is about 0.9. The increase in the C_D value on the large barge is believed to be due to the presence of the side walls which is expected to increase the force on the barge.

The results for C_D for the medium and large barge models are relatively independent of Re over the range of Reynolds numbers tested. The C_D results for the small barge without bilge keels indicate a general increase in C_D with a Re number, even over the relatively small range of Re numbers tested. At low values of Re, the value for C_D is approximately 0.7 which is similar to the results for the medium barge. It is believed that because of the size of the small barge, the flow around the barge is more three-dimensional than two-dimensional. For the small barge, the flow separation effect around the ends will affect a greater relative portion of the length of the barge. The three-dimensional flow around the barge will cause a larger pressure difference fore and aft (due to larger wake behind the structure) resulting in larger relative load. The effect of turbulence (and hence wake) will increase with the increasing velocity (i.e., flow). Hence, there is an increase in the C_D value with Re.

From the above discussion it is clear that of the three sizes tested in the tank, the most optimum size is the medium barge. It is the least susceptible to the side will effect and to the three-dimensional effect. Since the mean C_D values between the small and medium barges are approximately equal (about 0.7), the medium barge, $W/L = 5$, is considered to have a minimal effect from the walls of the tank.

It is also clear from these discussions that the blockage factor, f, as defined by Eq. 3, does not necessarily represent the true blockage effect from the sidewall. Seelig et al. (1992) demonstrated that for a given tank width the

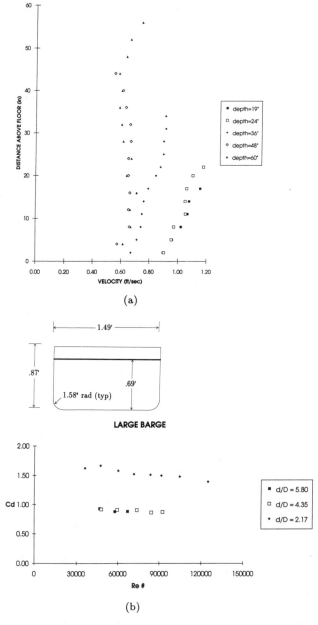

Fig. 17. Drag coefficients for barges in various water depths. Corresponding current profiles are shown on top.

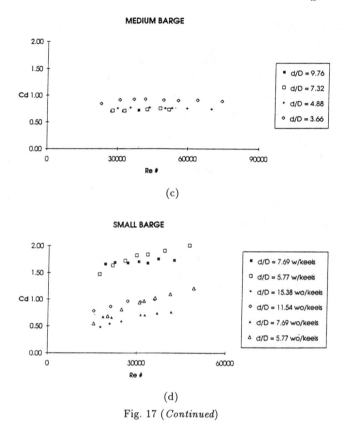

Fig. 17 (*Continued*)

correction factor works for shallow waters. The results of the tests described here further indicate that the effects of the tank walls and tank floor should be treated independently.

6.3. *Sheer current on a tanker*

Data generated during the tests of a T-2 tanker model in a current (Chakrabarti *et al.*, 1994) clearly demonstrate the effect of shear current on the lateral forces. Similar results are available from the field tests on a prototype T-2 tanker. The test was performed with the tanker model moored in a steady current. Two mooring lines equipped with load cells held the floating model in place, but the model was allowed to roll in the current. Uniform and shear currents were generated in the laboratory which overlapped the prototype values if Froude's law is followed. Various strengths of positive and negative shears were generated in deep and shallow water. The shear currents were generated by providing increasing levels of friction to the flow at prescribed elevations. This friction was achieved by placing a mesh, consisting of a polyethylene hardware cloth, in front of the flow straighteners.

6.4. Deep and shallow water results

Test results show that for the same test set-up, the shallow water lateral force is somewhat larger than the corresponding deep water force. For example at a current speed of 0.8 ft./sec. in uniform flow in beam seas, the deep water lateral force in the model was 2.5 pounds while the corresponding load in shallow water was 3.0 pounds.

In both shallow and deep water, shear in the current had a significant impact on the lateral force. Negative shear produced the highest loads while positive shear produced the lowest values. In deep water, the difference in lateral forces between the zero (uniform flow) and positive shear was found to be minimal (Fig. 18). This is confirmed by the measured heel angle. In fact, the trend in the heel angle corresponded with the lateral force behavior in all cases.

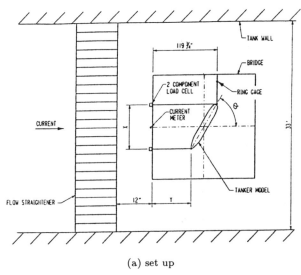

(a) set up

Fig. 18. Current load on a moored-floating (T2) tanker model in two orientations.

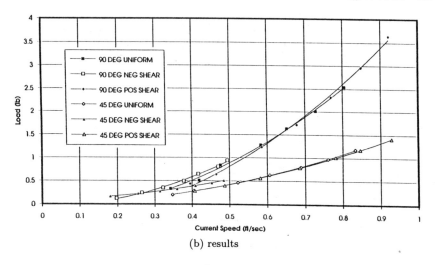

(b) results

Fig. 18. (*Continued*).

6.5. *Validity of model tests*

The lateral force coefficient and the yaw moment coefficient are calculated at each heading angle θ_c (0° for the head sea and 90° for the beam sea) from the formulas:

$$C_y(\theta_c) = \frac{Y_c(\theta_c)}{\frac{1}{2}\rho L_{pp} T U^2} \tag{5}$$

and

$$C_n(\theta_c) = \frac{N_c(\theta_c)}{\frac{1}{2}\rho L_{pp}^2 T U^2} \tag{6}$$

where Y_c and N_c are the mean values of the lateral load (down-tank load) and the yaw moment computed from the measured load cell readings, ρ is the mass density of water, L_{pp} is the tanker model length between perpendiculars (80.5 inches), T is the vessel draft (5.04 inches), and U is the mean current speed measured ahead of the model at its mid-draft.

The Reynolds number is defined as

$$\text{Re} = \frac{U(2T)}{\nu} \tag{7}$$

where ν is the kinematic viscosity of water (1×10^{-5} ft.^{2}/sec. for model and 1.4×10^{-5} ft.^{2}/sec. for full scale).

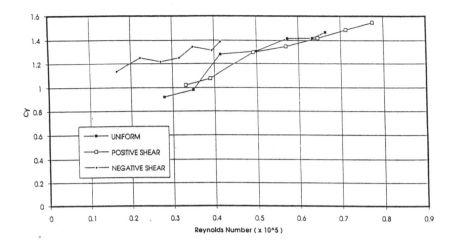

Fig. 19. Drag coefficient versus Reynolds number for a floating T2 tanker model in deep water.

The values of C_y versus Re for deep and shallow water cases are shown in Figs. 19 and 20 for the 90-degree orientation of the model. In deep water, the zero and positive shear produced similar values of C_y. However, the values of C_y in negative shear were decisively higher. In shallow water, the positive shear C_y was consistently lower than the uniform flow C_y while the negative shear C_y was substantially higher. For other tanker orientations, the observations were identical. These results are similar to the trends of C_y found in the field tests of the T-2 tanker (Palo, 1986), but the difference is more dramatic.

Additionally, in all cases the values of C_y tend to increase with the Reynolds number within its test range. In deep water as well as in shallow water, zero or positive shear, the slope is small. However, the negative shear in shallow water shows much higher slope in the C_y values. This observation can also be made with the prototype data which were at a much higher Reynolds number. Therefore, the model tests were representative of the prototype. The values of C_n did not exhibit any definite trend with the Reynolds number for either water depth. Qualitatively, the model results seem to verify the prototype trends. The increase in the lateral force coefficient with the Reynolds number does not agree with the OCIMF (1977) results. In the latter case, the Reynolds numbers were higher and at the highest Reynolds number (Re = 1.8×10^{-5})

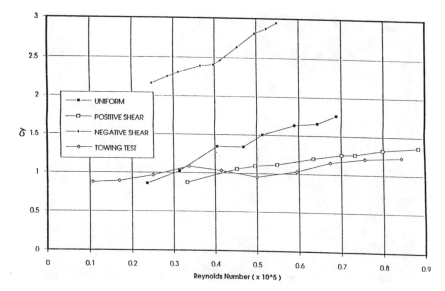

Fig. 20. Drag coefficient versus Reynolds number for a floating T2 tanker model in shallow water.

no vortex shedding phenomena were observed. In the present case, Re was less than 10^5 in all cases and visual observation showed increased vortex shedding activities with increasing Re. This may partly explain the increase in the lateral force coefficient with Re.

The shallow water plot (Fig. 20) includes a trace generated by towing the model in still water. While it was expected that the towing results would be comparable to the uniform current results, this was not the case. The C_y values determined from the towing tests were generally lower than those from the uniform current tests and showed only a small rise for increasing values of Re. The differences in C_y may be due to observed differences in the flow around the model in the two tests. Figure 21 shows two photographs of the flow around the model for the towing test and the uniform current test. The lower photo (towing test) shows evidence of waves being generated by the model as it moved through the water. The waves are not apparent in the current test photo. Also, the area of flow disturbance and vortex shedding behind the model appears to be larger and less separated from the model during the towing tests.

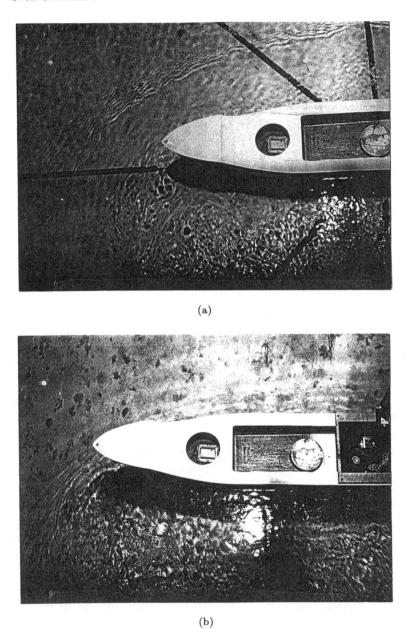

(a)

(b)

Fig. 21. Flow pattern around tanker model (a) in the presence of a uniform current and (b) while towed in still water.

6.6. Angular distribution of lateral loads

The tanker model was towed in still water to determine how the force coefficient C_y varies with the model's heading angle. Tests were made at towing speeds of 0.75 ft./sec. and 1.5 ft./sec. Headings of 0 to 180 degrees, at increments of 10 degrees, were tested for each of the two towing speeds. The bow-first heading was called zero degrees. The model was hinged at its top to a load cell mounted at the end of a vertical staff attached to the towing carriage. The hinge allowed the model to heel forward during towing. The heading angle was made adjustable using a lockable pivot between the tanker and the load cell.

The average force coefficient C_y was calculated for each test run and the results are plotted in Fig. 22. Also, the data was fitted with an equation of the form:

$$C_y = a + b \sin^c \theta_c \tag{8}$$

The coefficients a, b, and c were found to be 0.04, 0.99, and 2.54 respectively. These results lead to the approximation

$$C_y = \sin^{2.5} \theta_c . \tag{9}$$

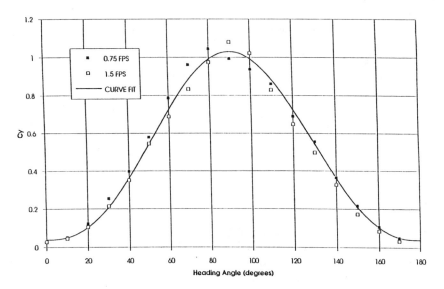

Fig. 22. Drag coefficient versus heading angle in tow of a T2 tanker model.

This is similar to the results of Palo (1986) except, in his case, the exponent on the sine term was given as 1.5 instead of 2.5. The coefficient a here (0.04) represents the force coefficient at 0 as well as 180 degree heading. The value of the coefficient is small compared to the maximum value of C_y; therefore, it may be neglected in the approximation given in Eq. 9.

Towing test data showed that the loads are very sensitive to the speeds near a 90-degree orientation. In fact, at a lower speed (0.75 ft./sec.), the maximum load (Fig. 22) occurred at 80 degrees. The measured loads in a current show a similar result. For a perfectly symmetric body, the lateral load should be maximum at 90 degrees. However, because of the difference in the tanker geometry at the bow and stern, the flow around them will be asymmetric which may cause a higher load at a slightly different angle. Similar asymmetry can be found in the results in the OCIMF report (their Fig. 8).

References

Bishop, R. E. D. and Y. Hassan (1964). The lift and drag forces on a circular cylinder oscillating in a flowing fluid. *Proc. Roy. Soc.* (Lond.) **277**:51–75.

Chakrabarti, S. K., A. R. Libby and P. Palo (1994). Small scale testing on current-induced forces on a moored tanker. *Oc. Eng.* (to be published).

Chakrabarti, S. K. and D. C. Cotter (1994). Tank wall effects on broadside current forces on barge models. *Oc. Eng.* **21**(5):489–497.

Chakrabarti, S. K., D. C. Cotter and P. Palo (1993). Shear current forces on a submerged cylinder. *Oc. Eng.* **20**(1):135–142.

Danish Hydraulic Institute (1986). Investigations and tests to determine hydrodynamic forces and moments on ships moored in a current. Vol. I & II, Phase I Report, CR 87.002.

Davies, M. E. (1986). The effects of turbulent shear flow on the critical Reynolds number of a circular cylinder. National Physical Laboratory, UK, NPL Report Marine Science R151.

Edwards, R. Y. (1985). Hydrodynamic forces on vessels stationed in a current. *Proc. 17th Annual Offshore Technology Conf.*, Houston, Texas, USA. Paper OTC 5032. 99–105.

Elder, J. W. (1959). Steady flow through non-uniform gauzes of arbitrary shape. *J. Fluid Mechanics* **23**(2):355–368.

Fellmann, M. (1992). Towing stability model experiments for YCV-19 series open lighter barge. David Taylor Model Basin CDNSWC Report DTRC-SHD 1375-01.

Griffin, O. M. (1985). Vortex shedding from bluff bodies in a shear flow: a review. *J. Fluids Engineering. ASME.* **107**:298–306.

Grosenbaugh, M. A. and D. R. Yoerger (1989). A full scale experimental study of the effect of shear current on the vortex-induced vibration of a long tow cable. *Proc. 8th Int. Offshore Mechanics and Arctic Engineering Conf.* Hague, The Netherlands. ASME. 295–302.

Hoerner, S. F. (1965). *Fluid-Dynamic Drag.* Published by the author.

Humphries, J. A. and D. H. Walker (1987). Vortex excited response of large scale cylinders in shear flow. *Proc. 6th Int. Offshore Mechanics and Arctic Engineering Conf.* Vol. 2, Houston, Texas, USA. ASME. 139–143.

Humphries, J. A. (1988). Comparison between theoretical predictions for vortex shedding in shear flow and experiments. *Proc. 7th Int. Conf. Offshore Mechanics and Arctic Engineering.* Vol. 2, Houston, Texas, USA. 203–209.

Jacobsen, V. and D. Jones (1987). Current loads on ships moored in shallow water. *Proc. 19th Annual Offshore Technology Conf.* OTC 5581. 259–268.

Kiya, M., H. Tamura, and M. Arie (1980). Vortex-shedding from a circular cylinder in moderate-Reynolds number shear-flow. *J. Fluid Mechanics* **101**:721–736.

Kotansky, D. R. (1966). The use of honeycomb for shear flow generation. *AIAA J.* **4**(8):1490–1491.

Kriebel, D. (1992). Viscous drag forces on moored ships in shallow water. US Naval Academy Report, Annapolis, Maryland, USA.

Kwok, L. C. (1989). Effect of current drag forces on two large bodies in close proximity. *Proc. 21st Annual Offshore Technology Conf.*, OTC 6177 Houston, Texas, USA. 609–620.

Le Méhauté, B. (1976). *An Introduction to Hydrodynamics and Water Waves.* Springer Verlag.

Mair, W. A. and P. K. Stansby (1975). Vortex wakes of bluff cylinders in a shear flow. *SIAM J. Applied Mathematics* **28**:519–540.

Masch, F. D. and W. L. Moore (1960). Drag forces in velocity gradient flow. *Proc. ASCE, J. of the Hydraulics Division* **86**(HY 7):1–11.

Maull, D. J. and R. A. Young (1974). Vortex shedding from a bluff body in a shear flow. *Flow-Induced Structural Vibrations*, ed. E. Naudascher. 717–729.

Maull, D. J. and R. A. Young (1973). Vortex shedding from bluff bodies in a shear flow. *J. Fluid Mechanics* **60**(2):401–409.

Maull, D. J. (1969). The wake characteristics of a bluff body in a shear flow. *AGARD Conf. Proc.* No. 48, Paper 16.

Oil Companies International Marine Forum (1977). Prediction of wind and current loads on VLCCs. London, UK.

Palo, P. A. (1983). Steady wind and current induced loads on moored vessels. *Proc. 15th Offshore Technology Conf.*, Houston, Texas, USA. 159–166.

Palo, P. A. (1986). Current-induced vessel forces and yaw moments from full-scale measurements. NCEL Technical Note N-1749.

Peltzer, R. D. and D. M. Rooney (1980). Effect of upstream shear and surface roughness on the vortex shedding patterns and pressure distributions around a circular cylinder in transitional Re flow. Virginia Polytechnic Institute and State University, Report No. VPI-Aero-110.

Peltzer, R. D. (1982). Vortex shedding from a vibrating cable with attached spherical bodies in a linear shear flow. Naval Research Laboratory Memorandum Report 4940.

Rooney, D. M. and R. D. Peltzer (1981). Pressure and vortex shedding patterns around a low aspect ratio cylinder in a sheared flow at transitional Reynolds numbers. *J. Fluids Eng. ASME* **103**:88–96.

Seelig, W., D. Kriebel and J. Headland (1992). Broadside current forces on moored ships. *Proc. Civil Engineering in the Oceans V*. College Station, Texas, USA. ASCE.

Shaw, T. L. and M. R. Starr (1972). Shear flows past a circular cylinder. *Proc. ASCE, J. Hydraulics Division* **98**(HY 3):461–473.

Stansby, P. K. (1976). The locking-on of vortex shedding due to the cross-stream vibration of circular cylinders in uniform and shear flows. *J. Fluid Mechanics* **74**:641–667.

Zedan, M. F., A. Seif and A. Shibl (1988). An investigation of vortex shedding from cylinders in shear flow. *Proc. 7th Int. Offshore Mechanics and Arctic Engineering Conf.* Vol. 2, Houston, Texas, USA. ASME. 235–243.

Symbols

a	Amplitude of forced motion
α	Energy coefficient
b	Basin width
B	Ship beam
β	Shear parameter
C_b	Bottom effect coefficient
C_{D_f}	Drag coefficient at frequency f
C_f	Correction factor
C_{L_f}	Lift coefficient at frequency f
C_n	Yaw moment coefficient
C_p	Pressure coefficient
C_s	Surface effect coefficient
C_y	Lateral force coefficient
D	Cylinder diameter
d	Water depth
δ	Logarithmic decrement in air
ε	Velocity gradient
f	Frequency or blockage factor
f^*	Frequency ratio
f_c	Frequency of forced oscillation
F_{I_f}	In-line force at frequency f
F_{L_f}	Lift force at frequency f
F_M	Measured force on ship beam
f_n	Structure natural frequency
f_o	Shedding frequency for fixed cylinder
F_r	Froude number
f_s	Shedding frequency in forced vibration in fluid-flow
g	Gravitational acceleration
H	Chamber height
k	Surface roughness
k_s	Roughness parameter
K_s	Stability parameter
L	Cylinder length
L_{pp}	Length between perpendiculars
λ	Slope of current shear
M_e	Total mass (including added mass) of structure per unit foot

N_c	Current-induced yaw moment about vessel midship
θ_c	Incident current angle (0 for head sea)
P	Local pressure
P_d	Dynamic pressure
P_o	Stagnation pressure
R_A	Aspect ratio
R_a	Forced amplitude ratio
R_B	Blockage ratio
R_E	Elevation ratio
Re	Reynolds number
R_y	Free amplitude ratio
ρ	Fluid density
S_c	Strouhal number for cylinder in forced vibration in fluid flow
S_e	Strouhal number in shear flow as a function of local velocity
S_o	Strouhal number for fixed cylinder
T	Ship draft
T_I	Turbulence intensity
U	Local velocity
U_{dev}	Deviation of velocity from mean local velocity
U_M	Mean velocity in shear or uniform current
U_{MAX}	Maximum velocity
U_∞	Free stream velocity
ν	Kinematic viscosity of water
V_r	Reduced velocity
W	Chamber width
x	Longitudinal axis
y	Vertical axis
Y_c	Current-induced lateral force (down-tank direction)
Y_o	Amplitude of free motion
Y_1	Distance from cylinder axis to free surface
Y_2	Distance from cylinder axis to bottom boundary